SOLVING INTERFERENCE PROBLEMS IN ELECTRONICS

SOLVING INTERFERENCE PROBLEMS IN ELECTRONICS

RALPH MORRISON

A Wiley-Interscience Publication
JOHN WILEY & SONS, INC.
New York / Chichester / Brisbane / Toronto / Singapore

Library of Congress Cataloging in Publication Data:

Morrison, Ralph.
　　Solving interference problems in electronics / Ralph Morrison.
　　　　p.　　cm.
　　"A Wiley-Interscience publication."
　　Includes index.
　　ISBN 0-471-12796-5 (cloth : acid-free)
　　1. Electronic apparatus and appliances—Design and construction.
　2. Shielding (Electricity)　3. Electromagnetic interference.
　　I. Title.
　　TK7836.M65　　1996
　　621.382′24–dc20　　　　　　　　　　　　　　　　95-12074

CONTENTS

3 SIGNAL AND POWER TRANSPORT 36

4 RADIATION AND FIELD COUPLING 55

5 CALCULATING TOOLS

10   PRINTED CIRCUITS   177

11   SCREEN ROOMS   187

12 ELECTROSTATIC DISCHARGE **191**

APPENDIX   INTERFERENCE CASES AND THEIR SOLUTIONS   **194**

INDEX   **201**

PREFACE

I can remember writing the preface to my book *Grounding and Shielding in Instrumentation* back in 1967. I was very fearful that my efforts would not sit well in the engineering community. Who is this character writing about something everybody knows all about? With many thanks to the publisher, John Wiley & Sons, the book is still in print. How to ground and shield is still a mystery to many.

Over the years my involvement in instrumentation continued. Instruments changed character, the semiconductor revolution made big changes to electronics, but the noise and interference processes were ever present. I found myself still teaching courses in grounding and shielding, but with a change of emphasis. The new technologies demanded performance well above 100 MHz. In teaching these courses I had to make more use of my physics background. I began to realize that I had used simple electric field theory to describe analog processes, and now I had to think in terms of the electromagnetic fields to work at higher frequencies. In effect, it was field theory that mattered at all parts of the spectrum.

Teaching is definitely a learning experience. Many things become clearer in the process of answering questions. Often the gestation period for understanding spans years. I realized that there were some definite holes in my understanding of power and how it influenced interference. I got out the *National Electrical Code* and started to read. That was not a simple task, because first I had to understand I was reading a different language. It really came home to me that this issue of electrical interference was also tangled up in "word interference." If you work from a different set of definitions than I do, then we have a difficult communications problem.

I did not become a power engineer by reading the code, but I now know why electrical engineers do not understand power engineers. To bridge this gap, Warren Lewis and I set out to write a book titled *Grounding and Shielding in Facilities.* John Wiley saw some sense in this effort and decided to publish our effort. Again it was obvious that circuit theory did not function as a tool for processes that extended over large distances. A building is not a circuit, and the distribution of power is also not a circuit. The problem was to explain phenomena without using circuit theory, to have it make sense, and to explain how interference was generated and coupled to the user's equipment. In the end it was very elementary physics that provided the means to connect these aspects of interference with the circuit world.

At present I give seminars and do consulting work. The seminars have changed over the years because I am always trying to provide better ways to describe the general issue of interference control. The buzz words in the engineering community are still *grounding* and *shielding.* These are the words that attract attention and get students to sign up for a course. It is a trick because grounding is usually *not* the way to solve interference problems. It is important to know what grounding is all about, and then we can talk about the real issues.

A circuit diagram does not show physical size, relative position, physical spacing, or interconnection order. It says nothing about parasitics or loop areas. The effects of long cables or of the power grid are not even suggested. The electrical nature of the facility is not represented. Yet with all this missing information the engineer is supposed to function. It is a deep mystery to most, and the easy way out is to ignore it all until there is a problem. Give an electronics engineer the tools of his trade and ask him to characterize a building, and he will suddenly give you a broad grin and walk away.

It is well recognized that a good high-frequency circuit works because it is a carefully crafted geometry. It does not radiate, nor is it susceptible to radiation, because of its layout (geometry). Good radiators work because of their geometry. The relationship between circuits and geometry is simply physics. It doesn't have to be very mathematical, but it is the core of the topic. This is the approach used in this book. It is not a book of rules but of ways to view the world so that the geometries that must be used are obvious.

As a young student I was told the silly joke that to be an engineer one had to know Ohm's law and that you could not push on a rope. In the new age let me substitute for the rope a different aspect of mechanical engineering. To calculate area, multiply length times width. With Ohm's law and loop area at hand most interference processes are controllable. In the book it takes several hundred pages to convince the reader that this is indeed true.

This book follows the material given by the author in his seminars. Students often ask for a text to follow the course, and this has not been possible. If you read this preface, then the publisher must have also agreed with the students that the book is needed.

RALPH MORRISON

Eureka, California
January 27, 1995

SOLVING
INTERFERENCE
PROBLEMS IN
ELECTRONICS

CHAPTER 1

POWER

1.1 INTRODUCTION

Almost all systems require the use of utility power. When the utility power is disconnected, all hum, system noise, and interference seems to disappear (so does the signal by the way). This often leads to the odd conclusion that the noise is brought in with the power and that the utility company and/or the wiring in the facility are to blame. This belief is so strong that engineers often specify special power distribution for new facilities, which can result in unnecessary costs. Specifying every known "good" approach in the hope that there will be few problems later is not good engineering.

Modifications to existing facilities are often made without regard to how the changes will actually solve a problem. This "shot-in-the-dark" approach is also not good engineering. Once the issues are examined it turns out to be rare that the power itself is the cause of difficulty. When designs are correct, special power treatment is usually not necessary.

Another often-mentioned culprit is poor grounding. One might hear that "the system is noisy because the 'ground' is not good enough." There is a suspicion that the utility ground is involved since it is located at some remote undefined spot. The engineer notes that when changes to grounding are made the noise tends to change. This observation leads to unfounded conclusions. Connections to earth are necessary for safety reasons, as will be pointed out. Once the issues are examined, earth connections or grounding are not the cause of most difficulty. Audio systems work just fine on aircraft, and they certainly are not earthed.

1.2   POWER GROUND

A good place to start a discussion of power is to discuss how and why the utility company grounds some of its conductors. It is necessary to understand the meanings of the words that are used. Unfortunately, power semantics and electronic jargon are not the same. A good set of power definitions is written into the *National Electrical Code* (NEC). It would be a very haphazard world if these words meant different things to different power engineers. Power grounding is done by the book, and the power engineer does not have to experiment with this grounding to make the system function. In most cases an electronics engineer is not prepared to converse with a power engineer on the topic of grounding, because a different set of definitions is used. Within the electrical or audio engineering societies there is little agreement on a set of grounding definitions. There is always some effort afoot to offer this assistance in terms of special dictionaries, but these definitions are usually viewed as an annoyance. Language is a powerful force, and usage, not a dictionary, dictates meaning (or ambiguity).

Utility power is supplied to most facilities and residences. It must be safe and certainly not cause a fire or a safety hazard. The NEC is published by the National Fire Protection Association and is used by various jurisdictions and turned into building codes which are then made law. An audio or instrumentation engineer rarely has to face the issues of law involving power, although this may come to pass. To a power engineer, grounding and wiring per code takes care of making a facility safe from fire, lightning, and electrical shock. This may not be absolutely correct, as common sense still must be applied. An engineer searching for a solution to a noise problem might want to lift some power ground connection. This is unsafe and illegal and completely unnecessary when the entire issue of noise and interference is understood.

1.3   GROUND—THE DEFINITION ISSUE

The NEC defines ground as a connection to earth or its equivalent. A circuit designer may consider a chassis a ground whether it is earthed or not. To understand power engineers and their language, words or word groupings that use the root word *ground* must all be examined. Here are a few definitions:

Ground. A conducting connection to earth or its equivalent.

Grounded Conductor. A current-carrying power conductor that is earthed at the service entrance. It is the neutral conductor in a three-phase system.

Grounding Conductor. The specific conductor that connects power conductors at the service entrance to earth.

Equipment Grounding Conductors. Conductors (bare or green wires) used to ground the conductive (but normally not current carrying) metal parts of electrical equipment, including racks, frames, conduit, and housings, at the service entrance.

Grounding Electrode System. All of the metal in a facility, including conduit, water pipes, building steel, and the earth itself.

In this book when the subject relates to power or power grounding, these power definitions will hold. In other areas the meaning should be evident from the context.

The engineering fraternity has a lot of trouble with the general problem of interference. This includes engineering in the fields of instrumentation, analog and digital design, and audio engineering. In an effort to design problem-free systems many of the design concepts are implied in the words that are used. Unfortunately using these invented terms does not solve problems. In the search for solutions, grounding is assumed to be the culprit, and various word pairs have been put into use. For example, engineers will use terms such as *clean ground, digital ground, analog ground, power ground, audio ground,* and so forth. It is possible to find hundreds of these word pairs in the literature, and none of them are defined in any textbook. Implied in each word pair is some specific grounding treatment that is supposed to solve problems. The solution details are left to the reader's imagination. This book avoids such terms. Unfortunately, it is impossible to write a book on this general subject without using the word *ground* with all of its communication difficulties. Words such as "dirt," "mud," "mire," and "silt" are not available to differentiate different grounds. These words have not made it into the engineering lexicon.

1.4 POWER GROUNDING

The power brought into a facility is grounded at the service entrance. In the United States this grounding requirement is an important part of the NEC. A full discussion of this requirement would take several chapters. There are a few exceptions to this grounding practice, but this is not the subject of this book. Grounding in the power sense means that the grounded conductor or neutral is brought to the facility and tied to earth at the service entrance. This involves a separate and distinct conductor of specified size that makes an ohmic connection to the earth as a conductor. The Code specifies how good a connection this should be. In general a connection below 25 ohms is required. This grounding arrangement is shown in Figure 1.1.

The origin of this grounding practice predates any issues of noise or interference. It may surprise many readers to learn that this service entrance grounding is present to protect the facility or residence against lightning. When lightning strikes the utility conductors, it is seeking a low-impedance path to earth. It

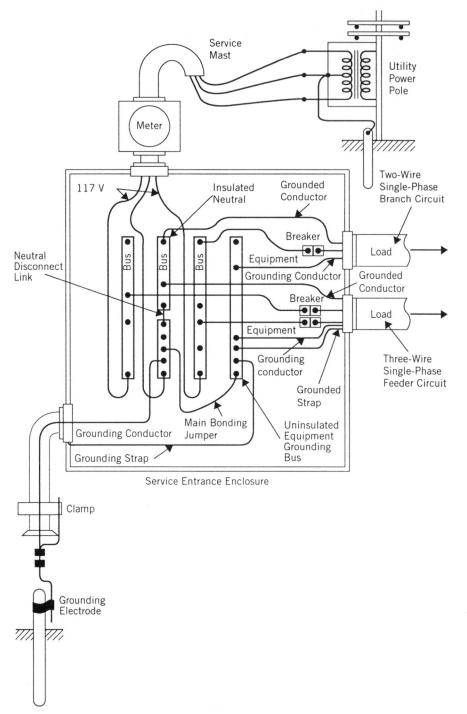

Figure 1.1   Grounding at a service entrance. (From *Grounding and Shielding in Facilities,* Ralph Morrison, Copyright © John Wiley & Sons, Inc., 1990. Reprinted by permission of John Wiley & Sons, Inc.)

is not a good idea to have the current path enter a building via the power conductors and get to earth via the plumbing. Any lightning path must be direct and low impedance, which implies that there should be no sharp bends in this path. When this direct path is provided at the service entrance, lightning will probably not enter the facility.

To illustrate the effect of inductance, assume a lightning pulse has a peak current of 100,000 amperes and a linear rise time of 2 microseconds. The voltage drop across an inductance of 1 microhenry is 50,000 volts. It only takes a few sharp bends to get to 1 microhenry. With voltages this high, arcing can alter the lightning path. The new path for the current flow is now undefined. If, for example, the new path involves rebars imbedded in concrete, heating at a poor bond could evaporate moisture, which could then tear apart the concrete.

The utility company often distributes power on utility poles. They want to keep lightning from blowing up transformers which could be along the lightning path. To limit their risks they often ground the neutral conductor at each pole along the distribution run. This conductor is often protected inside a wooden strip that runs up the side of the pole. This multiple grounding of the neutral conductor allows some of the neutral distribution current to flow in the earth. In a perfectly balanced ideal distribution system this current would be zero, but this rarely happens. The neutral current that flows in the earth tends to concentrate under the overhead conductors. Paths that are far from the line are too inductive and thus carry far less current.

The neutral or grounded conductor is grounded once at the service entrance for each facility and, by code, cannot be grounded a second time inside the facility. With all the power entry leads disconnected, there must be no ohmic path from power conductors to ground inside the facility. This is an extremely important point. This single-point ground is at the heart of the safety issue. All fault detection systems require that the return path for current for all loads be in prescribed conductors. If the grounded conductor shares a path through earth, then this concept is violated.

1.5 EARTH IMPEDANCES

The impedance levels involved in an earth path are often not appreciated. Soil resistances are usually measured at some low non-power-related frequency, such as 40 Hz. This is done to eliminate cathodic-type dc errors and stray power frequency pickup. A typical damp soil resistance is about 1000 ohm-cm. This means 1 ampere of current flowing uniformly across 1 cm^2 of soil will drop 1000 volts in 1 cm. The resistance is proportional to path length and inverse to path area. Thus 1 ampere flowing across a 1-m^2 surface over a 1-m length results in a voltage drop of 10 volts. If the measure were made over a cube 10 m on a side, the voltage drop would be 1 volt. This means that the resistance is 1 ohm. Obviously, when the current uses a large volume, the voltage gradient becomes quite small and thus the resistance becomes quite small.

For an earth resistance measurement, conductive rods must be driven into the soil. This means that the current flow in the earth must concentrate near the rods. This concentration of current limits the lowest value of resistance that can be obtained. This same concentration of current occurs when any point contact is made with the earth. Measurements taken under ideal conditions rarely fall below 1 or 2 ohms. To obtain a few ohms a deep well is used to carry the conductor several hundred feet deep. The well is chemically treated to increase the contact area. This discussion makes it very clear that a 25-ohm connection to earth as required by the NEC is a practical number and a good ground connection.

The average earth connection in damp soil through a single conductor is around 10 ohms. To obtain resistances that are 1000 times lower would require 1000 rods spaced uniformly over a rather large area. So far, this grounding concept has been limited to low frequencies, and therefore this measure of grounding is resistive. For power-related harmonic current flow at audio frequencies and at frequencies where radiation occurs, this resistance measure is inadequate. In most applications the earth impedance connection would probably measure hundreds of ohms above a few hundred cycles. Most engineers would not consider grounding a circuit through a resistor of several hundred ohms. For an earth connection the series resistance is simply at the other end of the circuit.

Suppose that there were two separate earth connections in one facility. Further suppose that the grounded conductor was connected to one earth and somehow the ungrounded conductor got connected to the other. What would happen? The earth would be heated by current flow. If the resistance of each earth connection were 10 ohms, the current flow for 120 volts would be 6 amperes. This is not going to trip any circuit breaker or blow any fuses. If someone touches conductors in contact with the two earths but not connected together, they would receive an electric shock, and this voltage is high enough to be lethal.

The lesson is simple. There can be one and only one grounding electrode system in a facility. This means that all grounded conductors must be bonded together and earthed. There can be multiple earth connections, but one of these connections must be available at the service entrance. Any attempt to construct a separate isolated earth system to "clean up" a system noise problem is unsafe and violates the NEC. As a matter of interest, the earth connection at the service entrance can no longer use the water pipe as the sole grounding electrode. Plastic couplers are often used by plumbers, and this interrupts the grounding path. The water pipe must be grounded, but a second acceptable grounding electrode must always be in place.

1.6 GROUNDING SEPARATED FACILITIES

Several more power grounding topics are worth discussing. The NEC allows two approaches when two buildings are considered to be one facility. The second facility can be treated as if it had its own service entrance. This means that the neutral or grounded conductor can be regrounded where power enters the second building. If there is going to be significant neutral or ground current flow, then some of this current will flow in the earth. If other metal conductors, such as gas conduit, connect the buildings together, then this is the path the current will take. There have been cases of a poor bond along a gas line, and heating at this bond has started a fire.

The second building can be treated as an extension of the first building, without regrounding the neutral or grounded conductor. The equipment grounding conductors must connect the two facilities together so that the fault protection systems can function.

The Code requires special treatment when there is more than one service entering a facility. This could be auxiliary power, backup power, or power at a different voltage or frequency. It is preferred that these systems enter physically next to each other to limit the impedance of the grounding tie since both systems must be bonded to the same grounding electrode system. The systems must be wired so that there is adequate fault protection for each system. This is certainly not the domain of the audio or instrumentation engineer, but he or she should be aware of this complication. Changes to the grounding scheme could violate the fault detection scheme, making the system unsafe. If a facility shows large voltage differences between various ground points, either the impedances are too high or there is significant neutral current flowing in the grounding electrode system. In either case the problem should be investigated by a competent power engineer.

1.7 EQUIPMENT GROUNDING

All metal housings, racks, and electrical equipment that might come in contact with an ungrounded power conductor must be grounded. Equipment includes conduit, outlet boxes, receptacles, panels, motor housings, transformer frames, and so forth. This grounding takes place through an equipment grounding conductor. If the conduit is the correct type, the Code allows the conduit itself to be the equipment grounding conductor. In most installations bare or green conductors are used. The Code specifies that the color green can only be used on equipment grounding conductors. For example, green stripes are not allowed for color coding any power conductor.

The concept is simple. If there is a fault to any metallic housing, the circuit must be interrupted. The fault circuit must provide a low-impedance path to guarantee a large current flow. This large current flow will force the interrupt to occur in a few cycles. *Note:* At the time of the fault, the housing is a shock hazard because it is at the full power potential. Fault currents should be many times the rating of the overcurrent protector to provide an immediate response. The impedance of the fault circuit includes both a resistive and an inductive component. If the fault path involves a large loop area, then the inductance can be high enough to limit the fault current. Under this condition the overcurrent protector may take a long time to react, and damage to equipment or to the facility can result. For this reason the Code requires that equipment grounding conductors run in the same conduit or cable tray that carries the power conductors.

To provide a low-impedance path, the equipment grounding conductors are all returned to the service entrance and bonded to the grounded conductors where they are earthed. An equipment grounding conductor may be multiply grounded (earthed) anywhere along its path. For example, the electrical conduit might be bolted to building steel or connected to a metal tank or touch another conduit. This is totally acceptable. It is incorrect to use these grounded metal structures for the only equipment ground connection. The reason relates to the fact that the fault circuit is then undefined and might be very inductive.

Equipment grounding conductors provide paths for many non-power-related currents. This point will be discussed later. This current is often called noise or interference. An engineer who considers this current flow undesirable may think he or she must break this current path, but this procedure is dangerous and violates the NEC. Disconnecting green wires or placing plastic couplers in electrical conduit places a facility and its personnel in jeopardy.

1.8 OSCILLOSCOPE GROUNDING

The electronics engineer faces a dilemma when he or she uses an oscilloscope. The manufacturer, by Code, must provide a three-pin power connector which grounds the oscilloscope case to the equipment grounding system (earth). Since most oscilloscopes are single-ended devices, the probe common is also connected to the oscilloscope case and thus earthed. When the probe is used to observe analog signals in a second grounded piece of equipment, a ground loop is created. Current flowing in this loop flows in the probe sheath and contaminates the signal. This ground loop can be avoided by not grounding the probe or by not grounding the oscilloscope case via the power cord. When the probe common is left disconnected, the signal seen by the oscilloscope can be quite noisy because the signal common return path involves a large undefined loop area, which picks up unwanted signals. When the oscilloscope is not grounded, it is technically unsafe. The probe common could be connected directly to the ungrounded power conductor, and anyone touching the case would receive a shock.

General instrumentation practice requires that the third-wire grounding pin be left ungrounded. The oscilloscope (or other equipment) is only grounded when the user makes an observation on a grounded circuit. This disconnect is usually provided by a "cheater plug," a three-wire to two-wire adapter.

Measurements made at frequencies above the so-called audio band may require a different approach to "grounding" the oscilloscope. This is discussed later.

1.9 ISOLATED GROUNDS

The word *isolated* as used here does not imply a floating ground or a separate or isolated earth connection. The NEC allows for a facility to be wired with special hardware so that the green wire for each circuit is brought back separately to the service entrance or an intermediate panel, where it may be spliced before returning to the entrance panel, without being connected to any other piece of equipment. The isolated ground is grounded at the service panel or it is connected to other equipment grounding conductors at the intermediate panel. As an example, in a power receptacle the third pin does not connect to the metal receptacle box but goes back separately to the service entrance. The receptacle box is still grounded but via the conduit or another green wire.

The intent of this wiring practice is to ensure that each piece of equipment has its own dedicated grounding conductor. When one equipment grounding conductor is shared by several pieces of equipment, the noise currents that flow can, in theory, cause noise cross-coupling.

If a user decides to cut the grounding pin or simply not use it, the equipment can be a safety hazard. If there is a fault to the case of the equipment, the equipment is "hot" and there will be no service interruption. A person touching the equipment could receive a fatal shock. Even when there is no fault condition, the ungrounded equipment will assume a potential determined by either power transformer or line filter capacitances. A user may still get a shock, but the current will be limited by a high series reactance. If a user is forced to cut the safety grounding pin to reduce noise, there is a systems problem that needs a solution. It would be unethical to request that a user cut this pin in order to make equipment function in a system.

The power line filters placed within a piece of equipment are usually referenced to the equipment ground (chassis). When an isolated ground system is used, the impedance in the noise current path is usually increased, thereby reducing filter effectiveness. When equipments are interconnected, a new set of problems can occur. This issue is treated in later chapters.

1.10 UNGROUNDED POWER

The NEC allows facilities to use ungrounded power. An example might be where a fault must be tolerated until equipment can be replaced, repaired, or sequentially turned off. The power design must provide for fault detection alarms, and a facilities engineer must be on call. These systems usually require special permission to operate.

Ungrounded power is used in a few instances. These power systems function quite adequately, but they do pose a new set of problems for equipment manufacturers who do their design while connected to a grounded system. These ungrounded systems provide for shock protection by requiring conductors to be in grounded conduit or adequately insulated. Computer manufacturers have found that their systems are often noisy or intermittent where ungrounded power is supplied.

The NEC allows ungrounded low-voltage power systems. This is not permission to use a single transformer to step utility voltages down to reduced levels. If there is a transformer fault, then the low-voltage system could become lethal. An ungrounded low-voltage system derived from a grounded low-voltage system would not fall under the rules of the NEC. Below 50 volts is considered low voltage.

The grounding or treatment of circuits within a piece of commercial hardware is not controlled by the NEC. Issues of safety or fire protection fall under the jurisdiction of organizations such as Underwriters Laboratories or its equivalent. The FCC is not interested in how a piece of equipment is built or grounded but only whether it radiates into space or conductively onto the power conductors. Manufacturers that intend to sell hardware to the European market must consider meeting new and more difficult regulations concerning this radiation. As equipment becomes more and more digital, designs to meet these new regulations become more complicated.

1.11 SEPARATELY DERIVED POWER

The secondary circuits of a distribution transformer or the circuits from an auxiliary power generator are considered to be separately derived sources of power. These derived sources of power must be grounded to the nearest point on the grounding electrode system of the facility. Just as with the service entrance, these power sources can only be grounded once. Since there can be but one grounding electrode system, all power grounds end up connected together. Attempting to provide a separate earth connection for separately derived power violates the NEC and is dangerous. The grounding of separately derived power is shown in Figure 1.2.

The advantage of providing a separate distribution transformer for critical loads is discussed in greater detail later. This advantage is maintained when the transformer is located near its load to keep the equipment grounding conductor and the neutral or grounded conductor runs short. For computer loads, special power centers are available, which include a transformer, circuit breakers, filters, and surge protection. These power centers can also be used effectively for other than computer loads.

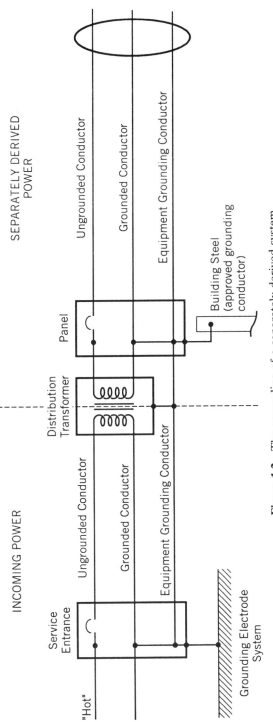

Figure 1.2 The grounding of a separately derived system.

1.12   THE k FACTOR IN TRANSFORMERS

Electronic loads are demanding higher and higher harmonic currents. In three-phase power distribution systems, the neutral current, in theory, can exceed the line current. This extra current has been known to burn out the neutral conductor while leaving the line conductors intact. High harmonic current must also be handled by any distribution transformer. Overheating can result unless the transformer is either oversized or designed to handle this type of load. This rating has been given the name "k" factor. It is the sum of terms involving the square of the harmonic number times the percentage of current load at that harmonic number. If 30% of the current is at the third harmonic and 20% is at the fifth harmonic, then the k factor is $0.5(1) + 0.3(9) + 0.2(25) = 8.2$. High harmonic current increases both the core and copper losses, which must be compensated by providing more iron and more copper.

1.13   PORTABLE GENERATORS

Earth grounding of portable generators is not required by the Code. The generator framework serves as the grounding electrode system. If the generator is located on a vehicle, the generator must be bonded to the framework of the vehicle. All loads must be located either on the generator or cord-and-plug connected through receptacles mounted on the generator.

1.14   GROUNDING OF VANS CARRYING ELECTRONIC EQUIPMENT

Vans operating from nearby building power must be treated like any other equipment load. The frame of the van must be considered equipment and connected to the "green wire" in the power cable. The neutral or grounded conductor may not be connected to the van frame or connected to earth.

If the van has its own distribution transformer, the secondary winding must be grounded to the frame of the van. The frame must still be connected to the equipment grounding conductor supplied in the connecting power cable. This point is also the connection for all equipment grounding conductors in the van.

A person stepping into the van has one foot on earth and one on a metal step. If the van is not grounded, a fault could put the van at a lethal power potential. Placing signs at the entrance suggesting that visitors jump to the first step is not satisfactory.

Electronic loads frequently have power filters that connect to the equipment grounds. Currents that flow in the equipment grounding conductors must find a path back to the power conductors. This connection is usually made at the service entry. For a typical van setup this may be a long distance indeed. When a distribution transformer is located in the van, this current need only return to the transformer. This transformer is considered a separately derived

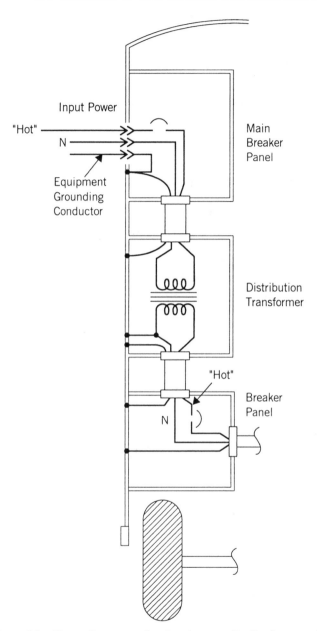

Figure 1.3   Grounding a van that has its own distribution transformer.

system, and the grounding of the secondary to the van frame is also the connection for all internal equipment grounding conductors. This grounding is shown in Figure 1.3.

1.15   POWER DISTRIBUTION IN EUROPE

In England the line voltage or "mains supply" is 240 volts rms, 50 Hz. In many parts of Europe the standard is 220 volts. Eventually this voltage will be standardized to 230 volts. In power cable (three-cored cable) the equipment grounding conductor is bare copper. Power is grounded (earthed) at each residence or facility via the structural steel. Traditionally this grounding was accomplished by an earth spike (grounding conductor). The issues of lightning protection and personnel safety are in effect treated just as in the United States.

Fusing is often provided at each receptacle based on the expected equipment load current. Typical fuse sizes are 2, 5, and 13 amps. Fusing is used to protect the wiring, not the equipment. In a facility, ground fault detection devices are installed, one for each floor. The power for each floor is supplied from a "ring main." In the United States these fault current devices are called ground fault interrupters (GFI) or ground fault current interrupters (GFCI). In England they are called earth leakage circuit breakers (ELCB). If excessive current flows in the equipment grounding path, the breaker interrupts the power.

In Europe the color code for equipment ground is green/yellow. "Live" or ungrounded conductors are brown, and neutrals or grounded conductors are blue. The equivalent of the NEC for England is the 16th edition of the *IEE Wiring Regulations* published in the United States.

CHAPTER 2

INTERFERENCE

2.1 INTERFERENCE PROCESSES—AN INTRODUCTION

Interference in electronic equipment is a constant source of difficulty for the design and systems engineer. The hardware may function adequately on the bench but may fail in some applications. Systems with long interconnects or many channels are often at risk. When the system is installed there is little that can be done inside of any piece of hardware to remedy a noise problem. If the interference seems to be power related, then there is little left but to try and modify the power system. Those who have faced these installation problems often insist that the power be modified, whether needed or not, before the installation proceeds.

Engineers have trouble drawing circuits that display the nature of interference. Even when the circuits can be drawn they provide little insight into how the interference processes work. Since grounding seems to be so important, it is often argued that a grounding schematic needs to be generated. This type of diagram is just a group of interconnecting lines plus some words. These lines provide little insight for analyzing a problem. The words often give the reader ambiguous information. Circuit theory is a powerful tool, but it does not fit a grounding schematic. Even with a schematic, circuit theory often does not fit.

An emphasis by schools on circuit theory is understandable because mathematics that fits all lumped-parameter problems is available. Unfortunately the steel structure in a building, the distribution of power, or the characteristics of a long cable are not classed as lumped-parameter systems. A long cable might be treated and analyzed as a distributed-parameter system. If the parameters are known, this class of problem can be handled by a special branch of circuit theory, and the treatment is quite mathematical.

What is a lumped-parameter system? It is simply a group of interconnected components that can store energy, dissipate energy, and provide gain. Usually there is an input signal, and the problem is to find the output signal. In such systems, energy is stored reactively in capacitors and inductors and dissipated in resistors. Gain can be provided by active devices such as transistors or vacuum tubes. The circuit's performance can be expressed as the ratio of voltages or currents or as the ratio of voltage to current or its reciprocal. Circuits of this nature can be analyzed by hand or modeled on a computer. The approach is direct, and given a well-defined problem every qualified engineer should get the same answer.

In a distributed system are capacitances and inductances that store energy, although not in specific components. A cable, for example, has capacitance and inductance per unit length. Transmission line theory analyzes idealized distributed-parameter systems and predicts input and output current or voltage ratios. Unfortunately the modeling is not very informative when it comes to interference processes. In practice, cables are very complex structures, and the best way to measure performance is by actual measurements. A building is an even more complex structure, and it does not remotely fit the ideas of a distributed-parameter system. It is interesting to ask the question, "How does one go about measuring a building?"

Circuit theory provides a shortcut method for handling a special set of physics problems. The behavior of circuits is not generally taught in a physics class. Yet all we know about electrical behavior stems from that branch of physics known as electromagnetism. When circuit theory fails to shed light on a problem, there is only one other place to turn—basic physics. This fact seems difficult to accept since most engineers remember vividly the difficulty of passing a course in electromagnetism. Physics, as taught in the universities, is highly mathematical, and its concepts can be obscured by the difficulty of using vector calculus and differential equations. It is not necessary to use this mathematics to understand basic principles, and it is in these principles that an understanding of interference can take place. Some mathematics is involved, but it is simple.

These arguments are presented so that the reader can appreciate why Chapter 2 discusses basic electromagnetic theory. This theory is an important next step in discussing the problems of interference, noise, and good design practice. The reader is reminded that circuit theory is still a vital tool, and it must be given its place.

2.2   THE *E* FIELD—AN INTRODUCTION TO VOLTAGE

To many the definition of the voltage involves a standard cell or a Josephson junction. This is the reference definition of 1 volt of electrical pressure, but it does not provide the meaning of voltage. To understand voltage, we must understand the concept of the electric field, which is basic for an understanding of all aspects of electrical behavior.

The story starts with the electric charge. Early experimenters found that when certain materials were rubbed together, electrical charges were moved from one body to another. By experimentation it was determined how it is possible to add charge or remove charge from various materials. The presence or absence of charge resulted in observable forces between the materials. It is an interesting physics experiment to add or subtract small charges from pith balls and see the balls attract or repel each other.

A charge is simply a group of electrons or protons. Electrons are the smallest units of negative charge. They surround the nucleus of every atom, which makes up all material. The number of electrons normally equals the number of protons, located in the center of the atom. Since each proton has a positive charge that exactly balances the negative charge on the electron, the atom is thus electrically neutral. This balance of charges is upset when electrons are removed from the surface atoms of a material, and the material is then said to assume a positive charge. Adding electrons provides a net negative charge. Note that protons cannot be moved around by rubbing the materials, because this would change the character of the material.

Bodies having the same charge polarity (i.e., positive or negative) repel each other, and bodies with opposite polarities attract each other. The force between charges can be measured by noting the mechanical forces that exist between charged bodies. These forces between charges are forces at a distance. The force at each point in space is ascribed to a field existing at that point. The nature of the force can be sensed by another charge, which has its own field. To map the field around a charged body, one uses a small test charge, small enough that it does not modify the nature of the field being measured. The force on the test charge measures the qualities of the field. This field is called an electric or *E* field, and because it has an intensity and direction at every point in space it is called a vector field.

The *E* field around a spherical conductor having a positive charge Q is shown in Figure 2.1. The lines leaving the surface of the sphere are lines of force. A positive test charge would experience a repelling force along the direction of the lines. The intensity of this force is proportional to the density of the lines. Near the surface of the sphere the force is greatest.

In a vacuum it is convenient to associate one line of force with each unit of charge. For the single sphere the geometry of the force field is simple. The lines of force are straight and extend to infinity. When two spheres are involved, the lines of force are curved and the field geometry depends on the charge on the second sphere. If the second sphere has an equal negative charge, the force field is shown in Figure 2.2. Note that all of the lines of force leaving the first sphere terminate on the second sphere, and none go to infinity. It is easy to see that in this representation the lines of force start on positive charges and terminate on negative charges.

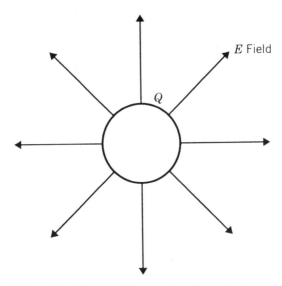

Figure 2.1   The E field around a charged sphere.

When a charged sphere is located above a large conducting plane, the field geometry near the sphere is as shown in Figure 2.3. Note that the field extends only to the surface of the plane. This plane of conducting material is often referred to as a ground plane. This description is appropriate because an infinite conducting plane somewhat parallels the characteristics of the earth. The lines of force that leave positive charges on the surface of the sphere terminate on negative charges on the surface of the conducting plane. An infinite conducting plane can be considered a source for any amount of positive or negative charge. In other words, any amount of charge is always available on the ground plane without doing any work.

Because a force attracts or repels a test charge, work must be done on the test charge in order to move it between two points. The maximum work per unit distance occurs when the test charge is moved along one of the lines of force. No work is done when the test charge is moved perpendicular to these lines. If the test charge is moved from a large distance to the surface of the sphere, then this additional charge can be accumulated on the sphere. Each increment of additional charge requires that work be added to the system. The process is very similar to raising buckets of water to the top of a water tower. The work to lift each bucket of water is stored as potential energy. This stored energy can be used to do work at a later time simply by allowing the water to fall back to earth.

The work done to add an increment of charge to the surface of the sphere must store energy. But where is this energy stored? Charge in itself does not store energy since the ground plane can supply any amount of charge and no

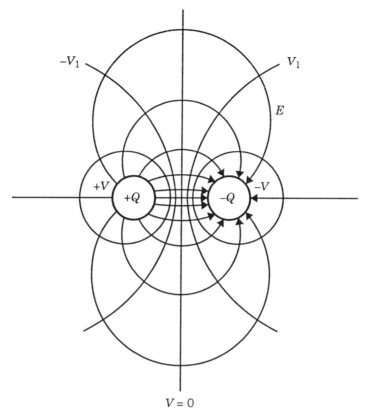

Figure 2.2 The *E* field around oppositely charged spheres.

work is involved. The sphere itself does not store any energy: it does not move or spin or contract. The only explanation that makes sense involves the electric field. It is easy to verify that doubling the charge on the sphere doubles the force on the test charge. The field is the only thing that changes. This means that the field itself must store the energy. It's a little bit like a coil spring that stores energy. When you look at a spring it is impossible to tell if it is compressed and storing energy. The sphere also looks the same whether it has a charge on its surface or not.

The idea that an invisible field stores energy in space is not an easy one to grasp. Once the concept is accepted, the rest is easy. In Figure 2.1 the field strength falls off as a function of the distance from the surface of the sphere. This must mean that the stored energy density is highest near the surface of the sphere. To get an idea of how to handle the electric field, we need a little mathematics. Experimentation shows that the force on a small test charge is directly proportional to the test charge value and the charge on the sphere,

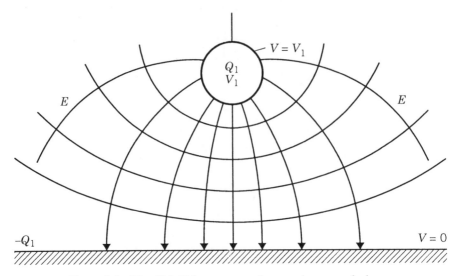

Figure 2.3 The E field between a sphere and a ground plane.

and is inversely proportional to the square of the distance from the center of the sphere. Stated mathematically:

$$F = \frac{k\,Q_1 Q_2}{R^2} \tag{2.1}$$

where F is the force on the test charge Q_1, Q_2 is the charge on the sphere, R is the distance to the center of the sphere, and k is a constant that includes the dielectric constant.

The work required to move a small test charge from infinity to the surface of the sphere can be found by simply integrating the product of force times distance summed over every increment of distance:

$$W = \int_{\infty}^{R} F \times ds = \frac{-kQ_1 Q_2}{R} \tag{2.2}$$

If Q_1 is set to unity, then the work per unit charge is

$$W = \frac{-kQ_2}{R} \tag{2.3}$$

This work is defined as voltage. The voltage at the surface of a sphere of radius R and carrying a charge Q is

$$V = \frac{-kQ}{R} \qquad (2.4)$$

where the potential at infinity is assumed to be zero. The force F in Equation (2.1) can now be written in terms of this voltage. Substituting Equation (2.4) into Equation (2.1), we obtain

$$F = \frac{-V}{R} = E \qquad (2.5)$$

In the MKS system of units (meters, kilograms, seconds), V and R have units of volts and meters. The force field (E field) is thus measured in units of volts per meter (V/m).

The E field has an intensity and direction at every point in space. The unit of meter does not imply that a meter of length is involved in the measure. To get a measure at a point, we must use a very small increment of distance. Thus, the E field is measured by noting the change in voltage over a short distance. Mathematically this is a derivative:

$$E = \frac{dV}{dR} \qquad (2.6)$$

This measure must be made in the space surrounding the conductor. On the conductor surface, charges are not moving, and thus there are no voltage differences. This also means that for the static case there is no E field inside any of the conductors.

2.3 VOLTAGE DIFFERENCES

The spheres and ground planes in the previous section were convenient. These simple geometries allowed us to discuss principles without undue complications. In practice, the conductors are circuit traces, cylinders, wires, coils, building steel, transformer iron, shields, conduit, and so forth. When voltages are placed on these geometries, E fields exist. A voltage difference implies that work must be done on the system to move a test charge across this voltage difference. This means that there must be an E field in the space between conductors.

Here are some ideas to consider:

1. If E fields exist in the space around conductors, energy must be stored in these fields.
2. The E field has lines of force that terminate on charges that must reside on the conductor's surface.

3. If the voltages are changed, the E field must change and charges must move in the conductors.

4. Energy levels cannot be changed in zero time because this requires infinite power. All field energy stored in the space between conductors must be accounted for in some manner when voltages change.

The term *potential* is often used interchangeably with voltage. A potential difference is simply a voltage difference. Note that there is no absolute zero of voltage. In a circuit or system one of the conductors is referenced as zero volts, and all voltages are measured with respect to this reference.

2.4 EQUIPOTENTIAL SURFACES

A conductor is an equipotential surface when charges are not moving in it. In Figure 2.3 most of the lines of force leave the sphere on the side of the ground plane. This means that most of the charge resides on this lower side. The charge distribution on a conductor does not need to be uniform for the conductor to be an equipotential surface.

It is possible to describe equipotential surfaces in space where there are no conductors. When a test charge is moved along this surface, no work is done on the test charge. This surface is perpendicular to the lines of force. Consider the simple sphere in Figure 2.4. The voltage at the surface of the sphere is 10 volts with zero voltage at infinity. Surfaces of equipotential are also spheres. If the radius of the charged sphere is 10 cm, the equipotential

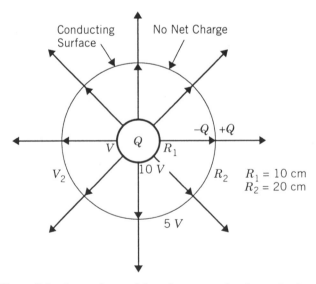

Figure 2.4 An equipotential surface around a charged sphere.

surface at 20 cm would be 5 volts, and at 40 cm would be 2.5 volts. As a matter of interest, the inside of the 10-cm charged sphere has no field. This is true if the inside of the sphere is a solid conductor or empty space.

If the equipotential surface at a radius of 20 cm were replaced by a very thin conducting sphere, nothing would change. Lines of force would terminate on the inside surface and leave on the outside surface. An experimenter could not tell the difference between measurements taken before or after the transformation. Since each line represents a unit of charge, it is easy to see that a charge $-Q$ exists on the inside surface of the 20-cm sphere and a charge Q exists on the outside surface. The net charge on our new surface is therefore zero.

With the second conducting sphere in place, let us remove the charge from the outer surface. Figure 2.5 shows a conductor grounding the 20-cm outer surface to a remote ground plane. By connecting these two conductors together they assume the same potential. If the ground plane is at zero potential, then so is the outer surface of the 20-cm sphere. Being at the same potential means that there is no work required to move a test charge from the ground plane to the outer surface of the sphere. This further means that there is no external E field and no charge on the outside surface of the sphere. A charge Q was obviously moved from the outside surface along the grounding conductor to the ground plane. Since the ground plane is infinite in scope, it can absorb this charge without any problem.

The 20-cm sphere now has a net charge of $-Q$. This charge is called an induction charge. The field on the inside of the 20-cm sphere remains unchanged. This is a very important observation. The field inside this enclosed conducting surface is unaffected by the presence or absence of an E field on the outside, regardless of the shape of the enclosure (conducting cylinder, box, etc.).

ELECTRIC FIELD INSIDE A GROUNDED CYLINDER

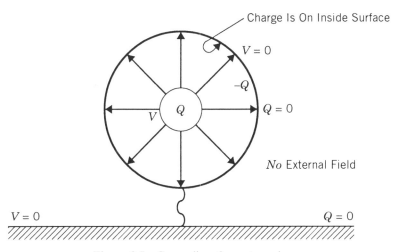

Figure 2.5 Grounding the outer sphere.

2.5 THE SHIELD CONCEPT

In the previous section it was shown that the E fields inside a conducting enclosure are not affected by E fields outside the enclosure. (We assumed a perfect enclosure without holes.) This idea is called electrostatic shielding. Limiting field entry is a powerful tool and one of the few devices available when interference must be blocked. A conducting enclosure is sometimes called a Faraday shield.

A conducting enclosure is a barrier against fields. Fields generated by internal electrical activity are kept inside and external fields are kept outside. Note that this form of shielding is independent of whether the enclosure is grounded; it works even if the enclosure is left floating. Shields are grounded for completely different reasons, which are covered later.

That different charge distributions can exist on opposite surfaces of the same conductor may come as a surprise. The charges that terminate a static E field must be on the surface. When an E field exists inside a conductor, there must be current flow. Any vertical E field component simply forces charges to go to the surface. In the static case this current does not exist.

Any steady currents that flow in a conductor require a constant tangential component of the E field. In copper this is the voltage drop per unit length. If the voltage drop in a piece of copper wire were 0.1 volt per meter, the E field inside the copper is also 0.1 volt per meter. At dc this tangential E field is uniform throughout the conductor. At higher frequencies skin effect takes place, and the current tends to concentrate near the surface. Skin effect is covered later.

When low-frequency currents flow in a cable shield, they use the shield's entire cross section. This means that the tangential E field that causes the current to flow must exist on the outside and inside surfaces of the shield. This fact says that a portion of the E field has penetrated the barrier. This effect can be lessened by using a shield with a lower resistance.

2.6 CAPACITANCE

The charge Q on the sphere in Figure 2.1 is associated with a voltage V. This voltage is the work required to move a unit charge from infinity to the surface of the sphere. The ratio of charge to voltage is called the self-capacitance of the system, as shown in Equation (2.7). The capacitance has units of farads when Q is

$$C = \frac{Q}{V} \tag{2.7}$$

given in coulombs and V is given in volts. In Figure 2.5 the voltage V is measured between the two spheres rather than between one surface and infinity. The field is fully contained between these two surfaces, and thus the energy stored in the field is also fully contained. This energy is exactly the same as the energy that could be stored in a commercial capacitor. To calculate this energy, solve for V in Equation (2.7). This is the work required to move an increment of charge dq between the two conductors. The total work W is simply the sum of all the work required to accumulate a charge Q. In integral form the total work is

$$W = \int_0^Q \frac{q}{C} \, dq = \frac{Q^2}{2C} \tag{2.8}$$

Substituting Q from Equation (2.7), we obtain

$$W = \tfrac{1}{2}CV^2 \tag{2.9}$$

Electric fields store energy. In a circuit sense this is reactive energy. It should be obvious that fields exist between all conductors when there are potential differences. These fields store energy, though not in a well-defined volume. At low frequencies this distributed energy storage is apt to be small compared with the energy stored in capacitors. As the frequency of interest rises, the capacitors in a circuit get smaller. The ratio of energy stored in the wiring to the energy stored in components thus gets larger.

Capacitance is a ratio of charge to voltage and is a property of conductor geometry. If the voltage is zero, the charge is zero and there is no field energy storage. The capacitance still exists, however.

2.7 MUTUAL CAPACITANCE

Consider the conductor geometry in Figure 2.6. A voltage V_1 is applied to a center conductor A. The outer conductor B has a small hole that allows some of the electric field to exit the enclosure. This leakage field terminates on an external conductor D. Hence, charge exists on this conductor since all lines of force must terminate on charge.

If the voltage V changes, then the charge Q must also change. This implies that there must be current flow in the grounding connection, since this is the only path for charge to take. The ratio between Q_3 and V_1 is called a mutual capacitance. The subscript notation C_{13} can be used. The ratio of Q_1 and V_1 is called a self-capacitance, and the notation C_{11} can be used. The sum of Q_2 and Q_3 equals Q_1.

Capacitance C_{13}, often called a leakage or parasitic capacitance, is not intentionally placed in a circuit. If the geometry in Figure 2.6 is cylindrical, this could be the leakage capacitance associated with a simple shielded cable.

INDUCED CURRENT FLOW

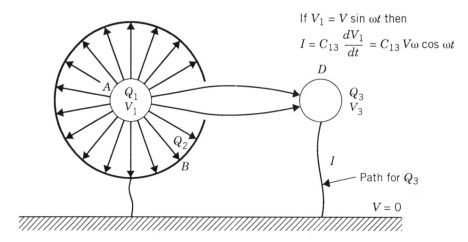

If $V_1 = V \sin \omega t$ then

$$I = C_{13}\frac{dV_1}{dt} = C_{13}\, V\omega \cos \omega t$$

Figure 2.6   A shield enclosure with a hole.

2.8   THE DISPLACEMENT FIELD

The E field we have used is not sufficient to handle the problems associated with different dielectrics. A dielectric is an insulator such as mica, mylar, oil, or paper. The presence of a dielectric enhances the capacitance, allowing more charge to be stored per volt. Dielectrics are also used in high-voltage applications to reduce arcing. The E field is a force field, but inside of a dielectric these forces are reduced by the dielectric constant, a property of the material.

To handle dielectrics, we introduce a second field parameter, called the displacement or D field. Earlier we identified one line of force with one unit of charge. Inside a dielectric the forces on a test charge are reduced, which means that the E field can no longer be associated with one unit of charge. To handle this difficulty, we associate the D field with charges and calculate the E field based on the dielectric constant. We write

$$E = \frac{D}{\varepsilon} \tag{2.10}$$

where ε is the dielectric constant.

To illustrate this idea, consider two parallel conducting plates forming a sandwich with two dielectrics, a simple capacitor (Figure 2.7). If the potential difference is 10 volts and the dielectric constants are 2 and 3, we know that the D field strength must be equal in both dielectrics. The problem is to find the E field in each dielectric. The E field in the first dielectric is $D/2$, in the second dielectric it is $D/3$. The voltage drop across the first dielectric is $D/2 \times 0.05$ m. The voltage drop across the second dielectric is $D/3 \times 0.05$ m. The sum of the two voltages must equal 10 volts, or

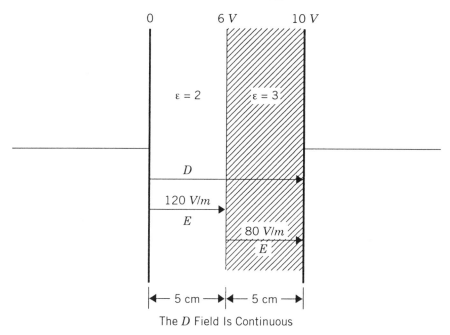

Figure 2.7   The two-dielectrics problem.

$$\frac{D}{2 \times 0.05\text{m}} + \frac{D}{3 \times 0.05\text{ m}} = 10 \text{ V} \qquad (2.11)$$

Solving for D, we find that it is 240 V/m. The E field in the first dielectric is $D/2$ or 120 V/m, and in the second dielectric it is 80 V/m. Obviously, the higher the dielectric constant the lower is the electric field strength.

The capacitance of this parallel plate system is proportional to the charge stored or to the value of D. With the dielectrics in place the D field is 240 V/m. If the dielectric constants were unity, D would be 100 V/m; thus the capacitance is increased by a factor of 2.4.

In a full treatment of the electric field a third vector is needed. Charges bound to the surface of a dielectric give rise to a field vector P, which is electric polarization per unit volume. In free space this vector does not exist.

2.9 THE MAGNETIC FIELD

A steady current flowing in a conductor creates a magnetic field. The field presence can be demonstrated by running the conductor through a piece of paper and sprinkling iron filings on the paper. The filings tend to form closed curves around the conductor. Magnetic forces align the filings. A tiny test compass can also be used to plot the shape of the magnetic field. The field at each point has intensity and direction and thus is a vector field. The shape of a magnetic field around a conductor is shown in Figure 2.8. A magnetic field exists around a permanent magnet as well as around a current-carrying conductor.

The magnetic field around a current-carrying conductor is called an H field. This H field at a point near a long straight conductor is directly proportional to the current and inversely proportional to the distance away from the center of the conductor. A second parallel conductor carrying current experiences a force when brought near the first conductor. Thus, the H field is a force field just as is the electric field. The direction of the field, the direction of the current, and the direction of the force are all at right angles to each other. For two parallel conductors carrying current in opposite directions, the force is in the direction to separate the conductors. For large currents these forces can be enormous. In a generator the currents flowing under fault conditions can destroy the windings.

Ampere's law states that the product of H field intensity times distance along a closed path equals the current encircled. If H varies along the path, then this product must be summed for every increment of distance. In integral form,

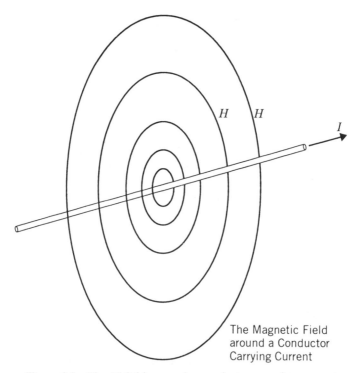

The Magnetic Field
around a Conductor
Carrying Current

Figure 2.8   The H field around a conductor carrying current.

$$\oint H \, dl = I \qquad (2.12)$$

If the path is a circle and H is constant, then

$$H = \frac{I}{2\pi R} \qquad (2.13)$$

The magnetic field around a coil of wire is proportional to the number of turns. In this geometry the path for determining H encircles the current n times, where n is the number of turns. In this geometry the lines of H are not circular and the calculation gets complicated. In the center of the solenoid the H field is approximately

$$H = \frac{In}{2\pi R} \qquad (2.14)$$

with flux lines directed along the length of the solenoid. The units of H are amperes per meter (A/m), as shown in Equations (2.12) and (2.13).

2.10 THE INDUCTION FIELD

Most magnetic behavior must be described in terms of two fields. This parallels the electric field case where the presence of a dielectric reduced the E field. A similar situation occurs when magnetic materials are involved.

Early experimenters discovered the principle of induction. When a loop of wire moved in a magnetic field, a voltage appeared across the loop. A voltage also appeared if the magnetic field was changed. The rules describing the induction of voltage when magnetic flux crosses a loop are known as Faraday's and Lenz's law. This induction is the basis of all motor and transformer action.

The magnetic field that induces voltage is known as the B field, thus the name induction field. This B field does not depend on material permeability. There is a parallel here with the electric field, where the D field does not depend on the material's dielectric constant. For the simplest case the B and H fields are parallel. Inside of magnetic materials having permeability, the H field is reduced. The ratio between B and H is

$$B = \mu H \qquad (2.15)$$

where μ is the permeability. Depending on the system of units used, μ can have dimension as well as a numerical value. Usually, H is expressed as amperes per meter, and B is in teslas (T). The tesla is 10^4 gauss (G), a more familiar unit. For B in teslas, $\mu = 4\pi 10^{-7}$.

Current flowing in a conductor creates a magnetic field. If the dimensions of the circuit are increased, the energy stored in the field becomes larger. The lines that represent the field are called lines of magnetic flux. The energy stored in a field is proportional to the total number of magnetic flux lines. Flux is simply the magnetic intensity times the area crossed by the flux lines. For all of the energy in a system to be considered, the area selected must intersect all of the lines of flux.

In a full treatment of the magnetic field a third vector is needed. This M vector is the magnetization per unit volume and relates to the magnetic properties of atoms. In free space this vector does not exist.

2.11   FARADAY'S LAW

The voltage induced in a loop of wire is proportional to the rate of change of induction flux crossing that loop and to the number of turns in the loop. Stated mathematically,

$$V = n\frac{d\phi}{dt} \tag{2.16}$$

The following interference problem illustrates the induction process. A circuit loop on a printed circuit (PC) board has an area of 100 cm². A power conductor 10 cm away carries a 10-A current. When the current is turned off it decays to zero in 0.1 microsecond (μs). The problem is to find the voltage induced in the loop. To use the preceding equations, we must convert all dimensions to meters. The H field is $I/2\pi R$ or 15.9 A/m. The B field is found by multiplying H by the permeability of free space. Thus, B is 2×10^{-7} T. The B flux ϕ is found by multiplying B by the area involved, which is 0.01 m². The flux is 2×10^{-9}. This flux changes from zero to maximum in 0.1 μs. Using this figure the voltage in the loop is 2 V. If this voltage is added to a 5-V logic signal, it could destroy an integrated circuit.

2.12   INDUCTANCE

A magnetic field stores energy. In the electric field, energy was stored in capacitance. In a similar way magnetic field energy is stored in inductance. Inductance is a geometric property, and there need not be current flow for there to be inductance.

Inductance is defined as the flux per unit current:

$$L = \frac{\phi}{I} \tag{2.17}$$

Note the parallel with capacitance, which is defined as the charge per unit voltage. It is difficult to measure flux directly, so other techniques must be used to measure inductance.

Faraday's law states that voltage depends on the rate of change of flux coupling to a circuit. With the definition of inductance, the voltage in Equation (2.16) becomes

$$V = \frac{dIL}{dt}$$

If L is not changing, this equation reduces to

$$V = L\frac{dI}{dt} \tag{2.18}$$

Engineers frequently use Equation (2.18) as the definition of inductance. Equation (2.17) is the proper definition, and Equation (2.18) permits a simple way to measure inductance. If a sinusoidal current flows in an inductance, the voltage is equal to

$$V = L\omega I \cos \omega t \tag{2.19}$$

where $\omega = 2\pi f$ is the radian frequency.

2.13 ENERGY STORAGE IN FIELDS

The energy stored in a capacitor was given in Equation (2.9). The idea that energy can be stored in a volume segment away from any conductors is, perhaps, new. To find the electric field energy in space, we consider, for convenience, two parallel equipotential square surfaces. Charges can be placed on these added surfaces such that the original field remains unchanged. This configuration is shown in Figure 2.9.

The E field is given by D/ε. The charge Q on the plate is simply DA, where A is the area of the plate. The work required to move a charge dq across a distance l is $El\,dq$. Since $E = D/\varepsilon$, the work summed over all charges is

$$W = \int_0^Q El\,dq = \frac{Q^2 l}{2A}$$

Substituting $D = QA$ and $E = D/\varepsilon$, we can rewrite the work W as

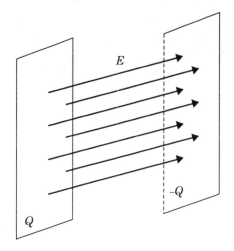

Figure 2.9   A segment of the E field in space.

$$W = \tfrac{1}{2}EDAl$$

in a vacuum,

$$W = \tfrac{1}{2}E^2v \tag{2.20}$$

where v is the volume of the field.

A similar argument can be made for a volume segment of the magnetic field. Unfortunately, no one has ever found the magnetic unipole that must be moved into the field to increase the current flow. To increase the energy in the field we must increase the magnetic flux in increments over the path length l. The work that must be done is proportional to the flux density present. In integral form,

$$W = \int Hl\,d\phi = \frac{\phi^2 l}{2A}$$

Substituting $\phi = BA$ and $B = \mu H$ gives

$$W = \tfrac{1}{2}BHv$$

in a vacuum,

$$W = \tfrac{1}{2}B^2v \tag{2.21}$$

where, again, v is the volume of the field.

2.14   WHY FIELD ENERGY IS SO IMPORTANT

Every voltage difference has a field associated with it. Every change is voltage requires that the E field change, which in turn requires charge to move. A moving charge is a current, and this produces a magnetic field. If the electric field is changing, there must always be magnetic field energy storage.

This simple idea is central to understanding electromagnetic processes, particularly interference. Every electrical activity results in an electric field and a magnetic field. At low frequencies the current flow is small and the magnetic field is weak, but the field is there. Only at dc can there be one static field without the other.

Both fields store energy, and both are needed to transport energy. When a voltage is sent to a circuit element, energy must be sent to fill the intermediate space with field. This viewpoint is not provided by circuit theory. It is this field that is related to crosstalk and interference coupling.

With circuit theory in mind, energy is stored in capacitors and inductors. With field theory in mind, energy is stored between conductors as well as in reactive components. When components are studied in detail, the basis is always field theory. A transistor requires an E field between the collector and emitter. A capacitor requires an E field to store energy. An inductor requires a B field to store energy. Even a resistor requires a potential difference or E field around its length before it can "resist." It is easy to argue that nothing electrical happens in this world without fields. The conductors in a circuit are there for one reason and one reason only—to transport fields to every component so that the components can function. In the process energy is stored between conductors and in components.

2.15   THE INDUCTOR

Inductors as circuit elements are used to store energy. They are used in passive filters to attenuate undesirable signals or select certain signals of interest. Inductors used at high frequencies can be wound in air or on ferrite-core materials. At power frequencies inductors often use transformer steel.

The magnetizing inductance found in a good transformer is high. This inductance is not intended to store energy. If the primary current contains a dc component, then the core may saturate. A magnetizing inductance of 10 henries (H) is not uncommon. A dc current of 100 milliamperes (mA) implies energy storage of 0.1 joules (J). This is a lot of energy storage, and a transformer is intended to transfer energy, not store it.

An examination of Equation (2.21) shows that, in any volume, stored field energy is proportional to the intensity of the B and H fields. If the flux path involves an air gap, then this is where the product of B and H is highest. If the permeability of the iron is 10,000, then the energy per unit length in air is 10,000 times that of iron. In typical inductors the gap is only a few thousandths of an inch long, whereas the total path length may be many inches. Inductors are designed so that the *B* flux is within the limits of the core material when maximum energy is stored. Again, energy is stored in space and not very well in materials. The purpose of the iron is to shape the field so that it is forced across a gap, since the energy is stored there.

Inductors that must store energy at high frequencies are not usually made from transformer iron because the core losses are too great. The preferred material is a powdered iron where the filler material between magnetic domains limits eddy currents, thus reducing core loss. The filler material can form a distributed gap, which can store energy between the domains of magnetic material. In some high-permeability ferrite-core designs (pot cores), actual air gaps are provided. Some cores provide a mechanical gap adjustment so that the inductance can be changed after the inductor is wound.

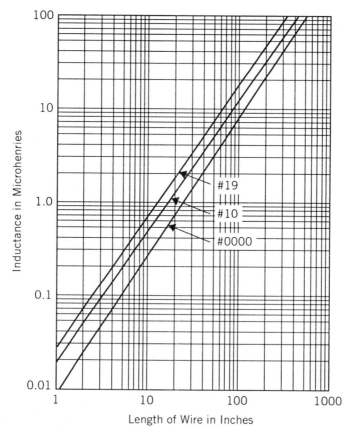

Figure 2.10 The inductance of straight conductors. (From Frederick E. Terman, *Radio Engineer's Handbook, First Edition,* McGraw-Hill, 1943.)

2.16 INDUCTANCE OF ISOLATED LEADS

All nonradiating circuits require a return path for all current flow. The concept of inductance has meaning when the energy stored in the inductance is returned to the circuit twice per cycle. Inductance can be calculated for various conductor geometries, including square or round loops, solenoids, or coaxial runs. It is not uncommon to consider the inductance of an isolated single conductor. The return circuit must exist, but it must be relatively far removed so that its fields are effectively separated.

The inductance per unit distance along the circumference of a large loop can be considered the inductance of a segment of straight wire. A plot of inductance for various lead diameters is given in Figure 2.10. Notice that the inductance depends strongly not on lead diameter but on length, because the magnetic flux depends on lead length, not lead diameter. A No. 19 lead and a No. 0000 wire have approximately the same inductance per meter. At frequencies above a few hundred kilohertz (kHz) the inductive reactance completely dominates conductor resistance. Later, when skin effect is discussed, this, too, will have a minor role in adding to the impedance.*

There is a strong desire to reduce unwanted potential differences by shorting them with a large conductor. The results are often disappointing. For the dimensions involved the connection can be hundreds of ohms of reactance. Even building steel can be ineffective, since it also has inductance per foot.

The problem is really the circuit view of the world. In large geometries the fields are only slightly modified by the conductors. A loop of wire cannot be used to short out a radio or television signal even if the loop happens to be building steel. It takes a different approach to eliminate the impact of an interfering field.

---

* E. B. Rosa and F. W. Grover, Formulas and tables of mutual and self inductance. *Bur. Standards* **8**(1) (1912).

CHAPTER 3

SIGNAL AND POWER TRANSPORT

3.1 INTRODUCTION

In many aspects of signal and power transport, circuit theory gives only part of the picture. At frequencies above 20 kHz where long distances are involved, circuit theory may be of little help. For digital processes where radiation is a distinct possibility, field theory ideas must be applied. In many situations power phenomena manage to couple to signal leads, resulting in a class of interference. These problems and their solutions are important to every engineer. This chapter discusses topics in which a combination of circuit theory and field theory is applied.

3.2 THE METAL BOX CONCEPT

When a simple circuit is contained in a metal box, all internal fields are kept inside and all external fields are kept outside. This statement must be qualified somewhat for low-frequency magnetic fields and for certain radiating sources. For the concepts to be discussed, the fields can be considered contained. In most applications, leads carrying power and signals must be brought in and out of the box. Holes must be provided for ventilation, and seams must be used to build the product. In other words, the ideal metal box never exists. The concept is useful, however, in illustrating some very simple ideas.

In Figure 3.1 a box with a circuit (battery operated) has one lead brought out to a connection on the ground plane. The circuit might be analog or digital, and the lead might be connected to another circuit. External electric

The Problem of Bringing One Lead out
of a Shielded Region

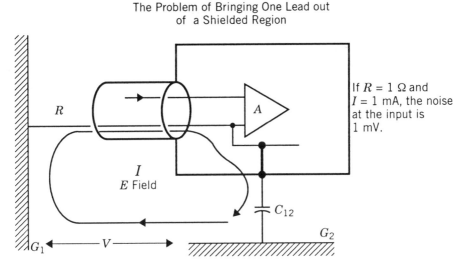

If $R = 1\ \Omega$ and
$I = 1$ mA, the noise
at the input is
1 mV.

Figure 3.1 A single lead exiting a circuit.

fields cause currents to flow on the surface of the box and now in the grounded
lead. The circuit diagram used to describe the source of this current flow is
usually a voltage source in series with a capacitance driving the box with
respect to the ground plane.

Assume the lead brought out of the box is an input signal common lead.
Small currents flowing in this lead develop small voltages, which are treated
as signal by the circuit. It only takes 1 mA flowing in 1 ohm to develop 1 mV
of signal. In many circuits a few microvolts (μV) are the permitted noise level.
This is a real and practical problem found every time a circuit connection
is made.

The first question is, should the box be connected to circuit common? The
answer is yes. If the circuit common is not connected to the box, the box
couples signals capacitively to the circuit and there is an undefined capacitive
feedback path from input to output. Every design engineer intuitively knows
that the circuit common should be connected to a nearby conductive enclosure
or there probably will be trouble.

Current flows in the signal common because a mistake has been made as
to where the box-to-common connection should be made. If a conducting
sheath is placed over the offending lead and the sheath is grounded where
the signal terminates, the problem goes away. This is shown in Figure 3.2. In
effect the box now has a different shape and extends over the common circuit
conductor. The ground connection to the circuit is now made to the outside
of the box. The sheath surrounding the signal conductor is recognized as a
simple shield. Currents now flow in the shield but not in the signal conductor.
The shield protects the signal by providing a path for unwanted current.

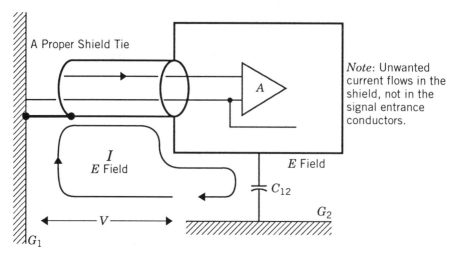

Figure 3.2 A proper connection to the shield.

3.3 THE TRANSFORMER ENTRY

Power for most circuits is supplied from a power transformer. The transformer allows the designer to adjust voltages and avoid direct connections to the power conductors. A transformer, however, provides an opportunity for external interference to circulate currents in signal or control conductors. The conductors supplying power are on the outside of the ideal box, and the secondary conductors of the transformer connect to the circuit common. Experimentation with this interface is not simple, because rebuilding or replacing transformers poses a serious obstacle.

The circuit symbol for a transformer is very misleading. For example, all of the turns of the primary coil are not adjacent to all of the turns of the secondary coil. In fact, the last turns of the primary are next to the first turns of the secondary. This is simply the way the transformer coils are wound in manufacture. There is a capacitance between the primary and secondary windings, and when the transformer is powered there is some average voltage in series with this capacitance. This voltage source causes current to flow in the grounded common connection, as shown in Figure 3.3. The last coils of wire of the primary may be connected to either the grounded or ungrounded power conductor. The voltage in series with the capacitance is obviously highest when the interfacing coil is connected to the ungrounded (hot) side of the line. This is the reason why reversing the power plug connection can sometimes reduce the "hum pickup" in a circuit.

The first idea that comes to mind is to complete the shielding formed by the metal box by connecting a transformer shield to the metal box. One end of the secondary coil is now next to this shield, and this coil voltage circulates current in the signal conductor through the coil-shield capacitance. A second

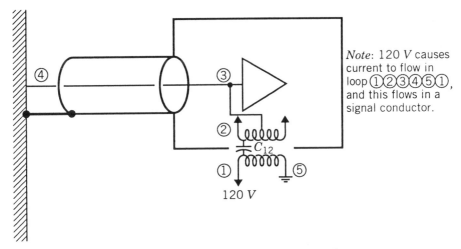

Figure 3.3   Current flow and the power transformer.

choice is to connect the shield to the circuit common. When this is done, the primary coil-to-shield capacitance in series with the primary coil voltage circulates current in the signal common. A third choice is to connect the shield to equipment ground. Again, this connection allows secondary voltages in series with ground potential differences to circulate current in the signal common. The conclusion one reaches is that there is no place to connect one shield in a power transformer to eliminate the flow of current in the signal common.

There is a benefit to connecting the transformer shield to the equipment grounding conductor. If there is a fault in the transformer between the primary coil and this shield, fault current will follow the proper low-impedance equipment ground path rather than the circuit common path. This is the safest shield connection. In instrumentation products where the case of the instruments cannot be equipment ground, this connection is advised. This shield is often referred to as a primary shield.

It is possible to build transformers where the secondary winding has a center tap balanced with respect to the transformer shield. Balancing reduces the equivalent voltage in series with the secondary-to-shield capacitance, and this limits parasitic current flow. Balancing is not generally available in commercial transformers.

A transformer shield is usually a single wrap of copper or aluminum foil. The wrap is insulated so that it does not form a shorted turn. This type of simple shield is also called a Faraday shield. Shields that wrap around an entire coil, the way a box might be wrapped, are called box shields. These shields are expensive because they require hand labor.

A measure of shield quality is its leakage capacitance. Correctly speaking, this capacitance is the mutual capacitance between the primary and secondary coils. Mutual capacitance is measured by establishing a ground reference plane. The transformer is removed from all of its connections and placed on this plane. The primary leads are all connected to an oscillator referenced to the ground plane. The secondary coils are all connected together and to the ground plane through a resistor. The shield is connected to the ground plane. The electric field that leaves the primary shield and terminates on the secondary coil stores energy in this mutual capacitance. If the shield were perfect, this capacitance would be zero. In the test circuit the oscillator lead must be carefully shielded or the measure will be invalid. This circuit is shown in Figure 3.4.

A simple shield in a typical 20-watt (W) transformer might have 5 picofarads (pF) of leakage capacitance. A box shield could have leakage capacitance as low as 0.01 pF. Measuring these small capacitances requires that the generator frequency be well above the power frequency. A 10-kHz signal is usually satisfactory. For measuring very small capacitances the detecting resistance should be at least 1 megohm. The input resistance of an attenuating oscilloscope probe can be used in lieu of a resistor. A 1.0-pF capacitance at 10 kHz has a reactance of 15.92 megohms. If the generator is set to 10 V/rms, the current flow is 0.628 μA rms. For a 1-megohm 10:1 probe this is 0.0628 V at the oscilloscope input. A 0.01-pF capacitance provides a signal of 0.628 mV of signal at the oscilloscope input. If the frequency is doubled, this signal level should also double.

It is easy to violate the shielding provided in a transformer. The primary leads must not be brought near the secondary leads unless these leads are inside of their own protecting shield. This procedure raises product cost, so the best way to protect these small capacitances is to pay careful attention to circuit layout. A power transformer installed so that the primary and secondary wiring enter the same receptacle obviously violates any shielding provided.

3.4   MULTIPLE TRANSFORMER SHIELDS

Currents that flow as a result of power transformer voltages, mutual capacitances, and ground potential differences can be reduced by using three shields in the transformer. The primary shield connects to equipment ground, the center shield connects to the metal box, and the third shield, which is next to the secondary coil, is connected to the circuit common. Such shielding is expensive, but, fortunately, rarely required. Power supplies that must float on a high impedance might be one application requiring three shields in the power transformer. Here the shields must limit the parasitic currents that would allow current to flow in the floating impedance.

A class of transformer called an isolation transformer is used in power distribution where sensitive electronic loads are involved. This transformer has multiple shields that are sometimes terminated internally. In many cases

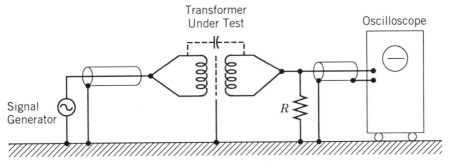

Figure 3.4 Testing a transformer shield.

isolation transformers are improperly applied. They are discussed in later chapters.

Switching-type power supplies use high-frequency transformers. These transformers often store magnetic field energy and operate at switching frequencies above 50 kHz. These rapidly changing voltages imply that the transformer is handling frequency content higher than 200 kHz. At these frequencies, any shielding adds coil capacitance by placing large reactive loads on the transformer, thus limiting its ability to function. For these applications, techniques other than shields must be used to limit or direct any high-frequency parasitic current flow.

3.5 THE GENERAL SOLUTION TO THE TRANSFORMER PROBLEM

It is difficult to keep a power transformer from circulating current in grounded leads that connect to a circuit. As indicated earlier, this can be a problem in low-level circuits where even microvolts of noise cannot be tolerated. In circuits with volts of signal and short signal runs, this current flow may not cause a problem. Examples where parasitic current flow can be tolerated might be the output of an amplifier or on lines carrying digital information.

This problem of conductive noise coupling is resolved when sensitive leads in question are not grounded. An example might be a microphone cable. If the circuits must be grounded, the problem may be resolved by using high-impedance differential input circuits. The current is limited by the high input impedance, which must be high enough to be effective since the source impedance usually involves a small series capacitance. The important point here is to recognize why the high input impedance is required. Understanding the phenomena allows the engineer to calculate the effect and prescribe a solution.

Most logic systems have no differential high-impedance interconnections, and the power transformers do circulate parasitic currents in interconnecting cables. When there is a voltage pulse on the power line, it has frequency content well above the power frequency fundamental. This voltage pulse causes current to flow in many of the interconnecting conductors. Any lead inductance can increase the voltage drop to the point where circuit damage can result. One way to limit these pulses is to use a line filter. If the impedance in the equipment ground path is high, a line filter may be ineffective. More will be said about line filters later.

3.6 TRANSFORMER ACTION

Some of the interference created by transformers requires an understanding of how transformers function. This section is not intended as a treatise on transformer design. Since transformers are a mystery to many, a discussion of basics is the only way to get started. Transformer action involves magnetic fields as discussed in the last chapter. To induce a voltage in a coil of wire, Faraday's law requires that there be a changing B field. A coil of wire with a changing voltage across its terminals implies that this B field exists. A nearby coil coupling to this field must have voltage across its terminals. Nowhere in these statements is there a requirement for transformer iron. So why is iron required in a power transformer?

The H, or magnetic, field that must exist in parallel with the B field requires current flow. In a typical power transformer where the primary has a few hundred turns of wire and there is no core, the current might have to be several hundred amperes to establish the H field. Since the H field is equal to the B field divided by the permeability factor, there is a way to reduce this current level. If the coils are wound on magnetic material, the H field level is lowered. A low value of H implies a much smaller current flow in the turns of wire, called magnetizing current. It has nothing to do with load current. This current must flow to establish the H field and, thus, the B, or induction, field. In an ideal transformer the magnetizing current is zero.

Faraday's law requires that the flux change if there is to be a voltage. The flux in question is simply the B field crossing the coil-coupling area A. If the voltage waveform is a sinusoid, the flux must be a cosine waveform. From Faraday's law the relationship between the B field and voltage is

$$v = An\frac{dB}{dt} \tag{3.1}$$

where n is the number of turns coupling to the B flux. If $v = V \sin \omega t$, solving for B gives

$$B = \frac{-Vn}{A\omega \cos \omega t} \tag{3.2}$$

where $\omega = 2\pi f$ is the radian frequency. When B is maximum, the voltage is at a zero crossing. Note that the maximum value of B increases when the voltage is raised or when the frequency is lowered. If B is expressed in peak gauss, V in rms volts, f in hertz, n is the number of turns, and A is the core area (cm^2), then Equation (3.1) reduces to

$$v = (4.44 \times 10^{-8})BknfA \tag{3.3}$$

where k is a stacking factor and usually has a value of about 0.9. Again, this equation says nothing about permeability, iron, or power level.

If the voltage is held constant, Equation (3.1) requires that dB/dt be a constant: In other words, B must be increasing at a steady rate. If the voltage is a square wave, the B field intensity is a triangular wave. As long as the voltage is positive, B increases. When the voltage reverses polarity, B begins to decrease. This flux-to-voltage relationship is shown in Figure 3.5.

The B and H curves for a typical transformer iron are shown in Figure 3.6. Note that the relationship between B and H is nonlinear. As the sinusoidal peak voltage is changed, the B/H curve that is traced changes. At lower voltages the maximum value of H tends to remain constant. At higher voltages the iron begins to saturate and H increases dramatically. It is easy to see that the ratio of B to H on some average basis varies with voltage level. One

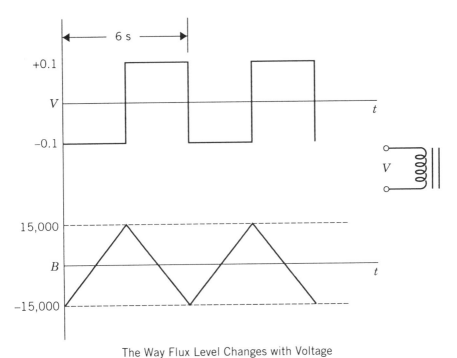

The Way Flux Level Changes with Voltage

Figure 3.5 The B/V relationship for square waves.

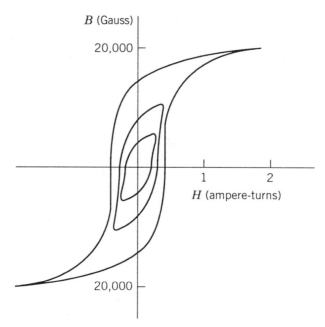

Figure 3.6 *B/H* curves for a typical transformer iron.

measure is to draw a line between the maximum *B/H* points and note its slope. This slope is a measure of average permeability. The slope for small voltage levels is called initial permeability. These *B/H* curves are also called hysteresis loops.

The increase in *H* for larger voltages is equivalent to saying that additional magnetizing current must be supplied. For sine wave voltages, maximum values of *H* occur when the voltage waveform goes through zero. If core saturation is suspected as a source of interference, the voltage waveform at a zero crossing should be examined. Excess current demands will show up as "glitches" in the voltage waveform.

Large power transformers require fewer turns per volt. The *H* field is proportional to current, to the number of turns, and to the length of the magnetic path. This means that a large transformer usually requires far more magnetizing current than does a smaller transformer. A 10-W transformer may have a no-load magnetizing current of 30 mA, whereas a 10-kW transformer may require more than 1 A of magnetizing current.

Transformer designs should allow for high-line-voltage applications. At maximum voltage the higher magnetizing current and resulting core loss can overheat the transformer. A transformer designed for 50-Hz operation will probably not saturate or overheat when operated at 60 Hz because *B* maximum is lower in the ratio 5/6. A 60-Hz design may saturate and overheat when operated at 50 Hz. This discussion does not consider heating due to load currents flowing in the windings.

3.7 LOAD CURRENTS IN A TRANSFORMER

The discussion has so far assumed an unloaded transformer. The voltage on the various transformer windings depended on the flux that is coupled and on the number of turns. If a load is placed on a winding, the current that flows obeys Ohm's law. The B field is defined by the voltage. The H field is defined by the B field and the permeability of the core. The net H field cannot change when load currents flow in turns of any secondary. This requires that the ampere-turns for all coils must sum to zero. Because the secondary load currents are fixed by Ohm's law, current must flow in the primary to balance H.

If there are 1000 primary turns and 100 secondary turns, 100 V on the primary places 10 V on the secondary coil. A 5-ohm secondary load requires 2 A to flow. This is 200 ampere-turns. To balance the H field, 0.2 A must flow in the primary coil. The resistance seen at the primary is 100/0.2, or 500 ohms. Note that this is 5 ohms times the square of the turns ratio.

3.8 LEAKAGE INDUCTANCE

Inductance is defined as the magnetic flux per unit current. In a transformer, leakage inductance is represented by that flux that does not couple the primary to the secondary coils. Leakage inductance acts as a reactance in series with the primary of the transformer. The physical space between coils, the space between turns, and the space not occupied by wire all contribute to leakage inductance. Transformers with cores of stacked laminations are apt to have more leakage inductance than transformers with torroidal cores. Torroidal-core transformers are usually smaller and more expensive.

Leakage inductance stores field energy in the space around a transformer. The energy in this field is a function of load current and magnetizing current. In large power transformers where load current can be hundreds of amperes, this leakage field can be significant. The leakage inductance of a transformer can be measured by shorting the secondary coils together and measuring the primary input impedance. The dc coil resistances must be considered in this measurement. Secondary impedances measured from the primary side appear as impedances multiplied by the turns ratio squared. If there are two secondaries, these impedances appear in parallel.

3.9 FIELD PROBLEMS AROUND TRANSFORMERS

The leakage fields near a power distribution transformer may pose a serious problem if certain precautions are not taken. These fields can induce significant currents into any conducting loops near the transformer. If the mounting supports are connected to building steel, induced currents can circulate throughout the steel in an entire facility. It is not uncommon for these current levels to exceed 20 A. This current can be confused with neutral current, which should not flow.

The cores used in open-cased transformers often have mounting holes. If these holes are used for mounting, the hardware must be insulated so that core flux does not thread a conductive loop. This flux exists anytime the transformer is excited, and does not depend on load.

The maximum B field in iron is usually kept below 18,000 G, whether the transformer is rated for 10 W or 10 kW. The B field near a small transformer falls off very rapidly. For there to be external fields comparable to the fields inside the core, the load current levels would have to be hundreds of amperes, a possibility in a loaded distribution transformer. When the current levels are a few hundred milliamperes and the magnetizing current is low, fields near a power transformer are usually not an issue.

Power transformers in electronic equipment are usually associated with power supplies using rectifiers and large electrolytic capacitors. Rectified currents flow in pulses near the peaks of voltage. These pulses of current flow in the leakage inductance of the power transformer and create rapidly changing fields. These fields are usually those that cause interference in nearby circuitry. Pulses occur at peak line voltage, whereas peaks of magnetizing current occur at line voltage zero crossings. In troubleshooting, the timing of interfering pulses can identify the nature of the difficulty. If the interference increases when the line voltage increases, the problem is usually related to magnetizing current.

The fields near a transformer can be shaped by placing a copper strap around the core. In some cases this technique can be used to limit interfering pickup. It is safer to reduce field coupling by limiting loop area in the signal circuit since transformer fields are not that easy to control. Also nearby equipment may introduce fields that cannot be canceled or manipulated. Small power transformers designed with flux levels below 8000 G can be mounted in microvolt circuits without causing interference as long as the signal loop areas are carefully controlled and the critical leads are kept more than a few inches from the core.

3.10   THE TRANSPORT OF SIGNAL AND POWER

Consider the simple circuit in Figure 3.7 consisting of a battery, a switch and a pair of conductors. The conductor pair has a capacitance and an inductance per unit length. At the moment the switch is closed, charges move to place energy into the first element of capacitance, and this moving charge stores energy in the first element of inductance. In the next increment of time more charge is moved to store energy in the second element of capacitance, and this moving charge stores energy in the second of inductance. As time proceeds, electric and magnetic energy is positioned along the line. This process takes place at a finite rate. It cannot happen in zero time because infinite

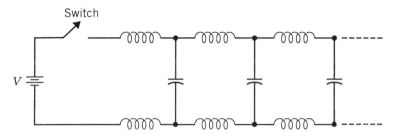

Figure 3.7 Signal transfer along a pair of wires.

power would be required. For this geometry the fields propagate along the line at about half the speed of light, or about 6 inches per nanosecond (in./ns). This is fast indeed, but for logic systems delays of nanoseconds in transporting signals can be an important consideration.

The energy transported per unit of time represents power. This power must be supplied by the battery and is simply the product of voltage and current. The energy transported on the line must come from the battery. The only transport mechanism that makes sense involves the two fields. It could also be stated that if both fields are present there must be energy flow.

The current that flows is steady since it supplies the same charge per unit time. The same current would flow if a resistive load were placed across the battery. In effect, this conducting line looks just like a resistor. If this resistance value were placed at the end of the conducting line, then, when the voltage reached this point, the same current would continue to flow. In effect, the line would appear to be infinite in extent. This resistance value is said to be the characteristic impedance of the line. The term *impedance* arises from transmission line theory, which treats the more general problem of energy transport in terms of sinusoids.

The battery in Figure 3.7 could be replaced with a signal generator and the ideas would remain unchanged. Energy is transported from the source of power to the load by fields whether the signal is dc or ac. This is certainly not the view taken by circuit theory, which pays little attention to the transport mechanisms. In power transmission the energy is carried in the fields, not in the conductors. Realizing that the purpose of wires or conductors is to direct where the fields go is a new perspective. The fields, not the conductors, carry the energy. Again, this is true at all frequencies, including dc. This viewpoint opens up an understanding of interference and interference coupling.

3.11 FIELD POWER OR POYNTING'S VECTOR

The fields around the conductors in Figure 3.7 provide some interesting information. The H field lines encircle each conductor, and the E field lines go between the conductors. A careful plot of these fields shows that the lines of

E and H are everywhere perpendicular to each other. Further, these lines are both perpendicular to the direction of energy transport. The E field intensity has units of volts per meter, and the H field intensity has units of amperes per meter. The product is volt-amperes per square meter at each point in space. If this product is correctly summed over an area, the power crossing that area can be evaluated.

The vector cross product of E and H at a point in space yields a power vector that points in the direction of power flow. This vector is called Poynting's vector. In Figure 3.8 it is shown at several points around the two conductors. The mathematics to handle the summation is not the subject of this book. Suffice it to say that the answer is V^2/R, where R is the characteristic impedance of the line.

The fields in Figure 3.8 follow a path directed by the conductors. Later, when radiation is discussed, the fields will not be so constrained but will be free to move in space away from any conductors. Both E and H fields are required for energy to move in space. The power density and direction at each point in space are given by Poynting's vector.

TRANSMISSION LINE — POWER FLOW

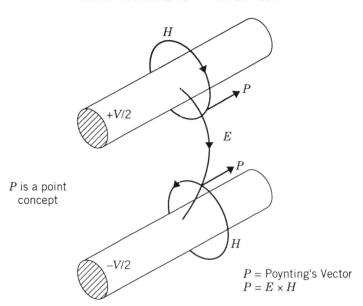

P is a point concept

P is perpendicular to E and H

P = Poynting's Vector
$P = E \times H$

$$P = \frac{\text{Volts} \times \text{Amperes}}{\text{Area}} = \frac{\text{Watts}}{\text{Area}}$$

Power flows in the space around the conductors

Figure 3.8   The power vector.

3.12 TRANSMISSION LINES

The pair of conductors in Figure 3.7 can be considered a simple transmission line. The field configuration in Figure 3.8 is symmetrical about a plane equidistant from the two conductors. This plane is an equipotential surface, and if a conducting plane is substituted for this surface the field pattern would not change. The potential between one conductor and the conducting plane is exactly half the potential between the two conductors. This field geometry is shown in Figure 3.9. If the second conductor is discarded, a new transmission line now exists between a conducting plane or ground plane and a single conductor. The characteristic impedance for this geometry is half that of the two-conductor case. The characteristic impedance for various conductor geometries is shown in Figure 3.10.

Every conductor pair can support the flow of field energy in both directions. This includes conduit, shield, earth, equipment ground, neutral, ribbon cable, trace, ground plane, or conductor in any combination. In most cases these conductor pairs are not held parallel, and they may terminate in undefined ways. When energy starts down one of these paths, the behavior may not be easy to describe. To get some idea of what might happen, we discuss a few special cases.

An open transmission line occurs when a conductor terminates in a relatively high impedance. Let's consider the battery and switch again. Energy proceeding down the line cannot be dissipated in an open circuit. Any energy in transit continues to arrive, and there is no way to stop it. The only mechanism that makes sense is for the energy to reflect and return to the source. The reflecting energy must cancel the current at the termination. For this to happen the reflecting wave must have the same voltage. The reflecting wave must be

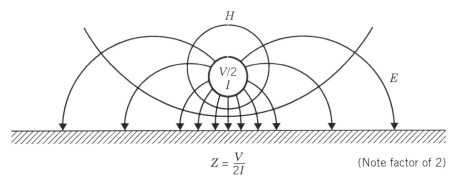

Transmission Line over a Ground Plane

$$Z = \frac{V}{2I}$$

(Note factor of 2)

The same field pattern
Field is concentrated under conductor

Figure 3.9 The field between a conductor and a ground plane.

L/l	Z (Ω)	H/h	Z (Ω)
1.1	53	0.6	37
1.5	115	1.0	79
2.0	158	2.0	124
2.5	188	2.5	138
3.0	212	3.0	149
4.0	248	4.0	166
5.0	275	5.0	180
10.0	359	10.0	221
30.0	491	30.0	287
100.0	636	100.0	359

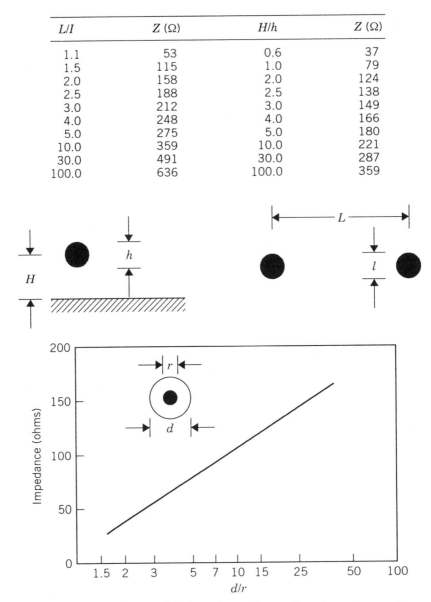

Figure 3.10   The characteristic impedance of several conductor geometries.

superposed on the first wave. This means that the voltage must double at the termination. Poynting's vector for the reflecting wave now points this second wave in the reverse direction. Now there are two waves in motion. Energy continues to leave the source while energy is being returned to the source.

This condition continues until the reflecting wave reaches the source. If the source is a zero-impedance voltage, a third wave must go forward with a negative voltage. To satisfy the direction of energy flow, this third wave must have a negative current. This process continues forever as long as there are no losses. Practically, the reflections eventually dissipate and the circuit assumes a steady state. In a circuit where the dimensions are a few feet, a dozen reflections would take 50 ns, a short time indeed.

Voltage doubling at an open termination can be used in various ways. In power distribution, lightning pulses often travel down the power line. The reflection at the end of a line can double an arriving pulse of voltage. The last user on the line may see this excess of voltage and suffer some damage. One solution is to add a final power support so that the reflection takes place at a remote point. If a flashover to earth takes place at this point, the pulse of energy is dissipated here rather than in a facility.

Energy proceeding down a transmission line terminated in a short circuit must be reflected because energy cannot be dissipated in a short circuit. This reflected wave must cancel the voltage at the termination. To return energy with a negative voltage, the current must remain positive. When this second wave returns to a zero-impedance source, a third wave must reflect forward, setting the voltage back to its original value. This wave must also have a positive current. The result is a tripling of source current. Whenever the wave makes a round trip, the current supplied from the source is incremented, but this increase in current cannot continue indefinitely for practical considerations. It is interesting to see how short-circuit current develops. It rises in a stairstep manner rather than in one giant step.

The ideas presented here are ideal. With a very fast oscilloscope these reflected waves can be demonstrated. In high-speed logic design these reflection processes must be given close attention. The elements of logic are nonlinear and have internal time delays, which further complicate the issues. Obviously if the correct signals are not present at clock time, then the design is inadequate or, at best, marginal.

The subject of transmission lines usually involves sinusoidal signals. High-frequency power is usually transmitted as a sinusoid. Source impedances, terminations, and reflections are a key issue for designers in this area. For sinusoids the reflected wave has a phase relationship that is a function of the transmission line length and frequency. If the line is terminated in a short or open circuit and the reflected signal returns a voltage equal to the original signal, then the line looks like an infinite impedance. If the reflected wave cancels the input voltage, the line looks like a short circuit. If the line is properly terminated, no energy returns. For an open- or short-circuit termination, an ideal line cannot dissipate any energy. For the general case a short or open termination makes the line look reactive, either capacitive or inductive. Again, this depends on line length and frequency.

3.13 COAXIAL TRANSMISSION

Examples of open-wire transmission lines may be found as traces on printed circuit boards or in ribbon cables. Here the signals are intentionally processed over parallel conductors as a matter of convenience. Most of the other open transmission paths that carry energy are a part of the general interference process. The fields carried by any open conductors, in theory, extend to infinity, and for this reason there is always some radiation and external coupling.

If the ground plane in Figure 3.9 is wrapped around the single conductor, the field is fully contained. Transmission between these two conductors is considered coaxial. The important point is that all of the signal current carried on the center conductor returns via the sheath. Ideally, external fields do not penetrate the outer sheath and the fields contained inside the sheath do not escape. The outer sheath of a coaxial cable can be used as a shield, but this is a different concept.

Coaxial transmission lines have the same characteristics as open transmission lines. See Figure 3.10 for the characteristic impedance of this geometry. The energy transported in a coaxial cable requires electric and magnetic fields. The same reflections occur for shorted or open terminations. Two-conductor coax is often used to transport high current levels. When the currents flow in the sheath, some of the energy can escape if the current penetrates to the outer surface. If kilowatts are involved, even a small power loss is undesirable.

3.14 DIRECTING FIELDS

The utility power we use is brought to us by conductors that direct the fields. As the frequency is raised, it gets more and more difficult to control and contain the fields and transport power. Long-distance power transport at 400 Hz is impractical for many reasons. One reason involves the fact that energy leaves the path provided by the conductors every time there is a turn or twist. Power must often be transported above 1 MHz, which can be done with coaxial cable. Examples might be radio or television transmission. Here the fields are contained in the coaxial geometry provided. At every anomaly, some of the energy may reflect, but at least it is contained. This field containment is an extension of the metal box idea discussed earlier.

Nature is always consistent. One of her traits is to find the easiest way to do something, another way of saying that water always runs downhill. The configuration taken always stores the least energy. The path taken by field energy follows the same idea. It is always trying to run down that hill. Sometimes this is difficult to see, but it is nonetheless true.

When field energy arrives at a group of conductors, some of the energy follows the conductors, since this is equivalent to doing the easiest thing. Once captured by the conductors, energy is on its way into any available box or circuit. It is important to recognize that nature is blind to labels, color codes, and our intent. A conductor pair of any sort is fair game for transmission, and this transmission can go both ways. Energy enters output leads as well as input leads. Energy travels between shields even if they are not intended for transport. We must have an open mind and expect any and all processes to take place.

3.15 MATCHING IMPEDANCES

A power source with a finite source impedance delivers maximum power to a load if the load impedance matches the source impedance. If the source impedance is zero, this idea of matching impedances must be reconsidered.

When a voltage source is indicated in a circuit, it is considered to be a zero-impedance source, which means it can deliver any level of current without changing its value. With feedback techniques it is possible to build a near-perfect voltage source over a finite range of current and frequency. In circuit testing, if the voltage is always adjusted to the same level as the test proceeds, it functions as an ideal voltage source.

When a transmission line is driven from a zero-impedance source, the energy transmitted is all dissipated in a matching terminating impedance. If the termination is a mismatch, some energy is reflected back toward the source. If the source has zero impedance, a second reflection must occur, sending this energy forward again. If the source impedance matches the characteristic impedance of the cable, this second reflection does not take place.

Adding a matching source resistor to a voltage source to drive a terminated transmission line reduces the terminating voltage by a factor of 2, impractical in many situations. Loss of signal level is not acceptable, especially if the signal is a logic level. In instrumentation where accuracies of 0.1% are required, an added resistor reduces full scale by a factor of 2, adds another accuracy factor, and involves the cable resistance.

When interference is coupled to a transmission line, matching impedances allow unwanted energy to be absorbed without reflection. If this is an important consideration, the disadvantages presented must be accepted. In blast study instrumentation, for example, reflections can disturb the data being measured, and matching impedances are important.

Long, unterminated shielded cables driven from a voltage source are rather common in analog work. A termination resistor forms an attenuator with the series cable resistance, and this affects accuracy. Depending on cable length, an unterminated cable usually has a peak in frequency response that is often not considered. One way to control this peaking is to terminate the cable in a series *RC* circuit. The capacitor blocks termination at low frequencies and allows termination at frequencies where peaking occurs. This *RC* circuit could be called a fractional termination. The value of *R* can be the characteristic impedance, and *C* can be chosen empirically to optimize the desired frequency response characteristic.

It is common practice to look at an unterminated cable as a reactive load. If the cable capacitance is 30 pF/ft, 1000 ft of cable has a capacitance of 0.03 μF. Signals would take about 4 μs to make the round-trip along this cable. This is one cycle at a frequency of 250 kHz. At this frequency the driving signal source sees an in-phase reflection, which is an open circuit. Obviously this cable cannot look like a simple capacitance to a driving amplifier. In fact, above this frequency the cable appears to be inductive.

The stability of amplifiers with capacitive loads is an important topic. It is proper to measure amplifier stability over a wide range of reactive loads. It is incorrect, however, to assume that a long cable looks like a simple capacitor. It is also incorrect to assume that an unterminated cable will have a flat frequency response.

3.16 THE REAL TRANSMISSION LINE

The distributed-parameter model of a transmission line assumes that the fields are confined to incremental inductances and capacitances. The analysis proceeds under the assumption that the energy stored in these elements all returns to the circuit. This is not completely true because some small part of the transported energy is actually radiated and never gets to the termination. This radiation is greatly limited when the transmission line is coaxial cable, since the fields are almost fully contained. For open conductors the fields extend to infinity.

At high frequencies, signal currents do not fully penetrate the conductors. This skin effect obviously impacts the nature of transmission. Skin effect is discussed in Chapter 4. A full treatment of transmission lines is thus very complex. The assumptions are necessary to get an idea of how signals and power are transported. Solving the bigger problem can easily obscure the basic ideas. These subtler aspects do occur and can affect the design of printed circuit boards, for example.

CHAPTER 4

RADIATION AND FIELD COUPLING

4.1 INTRODUCTION

Every voltage requires an E field, every current requires an H field. If one field changes, the other field must be present. The transport of signal or power requires the presence of E and H fields. In circuit analysis, field energy belongs to the circuit elements and stays in the circuit. Stored energy can be dissipated in resistances in the circuit or in loads presented to the circuit. In this ideal world, energy is brought in via the power conductors, manipulated, and sent to various loads. At no time does the energy leave the confines of the circuit except as heated air or, perhaps, light.

When circuit theory is not the view of the world, there is field energy stored between conductors. It takes time for fields to propagate between elements in this intermediate space. This process is no different than a transmission line except that some energy is not directed along specific conductors. When energy propagates into space, it travels at 300 m/μs. When the voltage and current return to zero, the fields must also return to zero. The energy must return to the circuit or continue its journey outward. Any lost field energy is said to be radiated.

For a circuit capacitance or inductance, all of the field energy is segregated and stored and returned to the circuit. For a sinusoidal steady-state analysis this give-and-take occurs once in each half-cycle. For energy that leaves the circuit, how are the two fields going to return to the circuit at the one point of origin? Obviously, this is not going to happen. When energy leaves the confines of a circuit, the story gets complex. A new viewpoint needs to be taken, one that is not too mathematical and yet gives some insight into what can be expected.

Most analog circuit designers should not have to worry about radiation. If their circuitry is imbedded in digital logic, the radiation from this logic can be quite significant. If radiated energy enters on input or output leads, it can create overload and distortion. Analog designers now have a nastier world to live with than they had 20 years ago. The problems they face extend over a much wider frequency spectrum than the old 10 Hz to 20 kHz. Today they must consider dc and be aware of signals that extend beyond 300 MHz. That is a lot to expect.

4.2 THE ELECTROMAGNETIC FIELD NEAR A SMALL DIPOLE

A dipole is a conducting geometry that can be used for receiving or transmitting electromagnetic energy. It takes the form of a conducting rod separated at its midpoint and connected to a transmission line. Energy confined by the transmission line is released by the dipole as radiation. Similarly, energy coupled from an external field is directed back along the transmission line to a receiver, for example. In most applications the intentional transmission of field energy involves a carrier signal usually above 1 MHz. To be effective as a radiator, the dipole dimensions should approach one half-wavelength at the transmitting frequency. A half-dipole consisting of a single rod is often placed perpendicular to a ground plane. This is the standard "whip antenna" found in so many applications. The conductor need not be perpendicular to the ground plane to radiate energy or couple to other external fields.

The analysis of field strength around a practical dipole is a formidable task. To get any picture at all, consider a very simple model. This model can be a small dipole element (a short piece of wire) with no diameter and a uniform current flow at one frequency. By putting together a group of these elements, we can approximate the characteristics of a practical antenna.

The fields around a simple dipole consist of terms that represent real and reactive energy flow. Near the dipole the E and H fields are roughly given by terms such as

$$E = k_1\left(\frac{\lambda}{2\pi r}\right)^3 + k_2\left(\frac{\lambda}{2\pi r}\right)^2 + k_3\left(\frac{\lambda}{2\pi r}\right) \tag{4.1}$$

$$H = k_4\left(\frac{\lambda}{2\pi r}\right)^2 + k_5\left(\frac{\lambda}{2\pi r}\right) \tag{4.2}$$

where λ is the wavelength and r is the distance from the dipole. When r is large, the last terms of Equations (4.1) and (4.2) dominate. Both E and H fields fall off linearly with distance, and the ratio of E to H is constant. When $r < \frac{\lambda}{2\pi}$, the first terms in Equations (4.1) and (4.2) dominate and the ratio of E to H increases as r decreases.

The distance $\lambda/2\pi$ is called the near-field, far-field interface distance. The field inside this distance is called a near electric field as the E field dominates. At distances beyond $\lambda/2\pi$ the field is called a far field. In the far field both E and H fall off linearly with distance, not as the square of distance. The far field is also described as a plane wave.

4.3 THE ELECTROMAGNETIC FIELD NEAR A SMALL LOOP

A loop of wire is an antenna and can be used for transmission and reception. Small conductive loops carrying current are found in most circuits, and these loops are all possible radiators. The E and H fields near a small loop are given by

$$E = k_1\left(\frac{\lambda}{2\pi r}\right)^2 + k_2\left(\frac{\lambda}{2\pi r}\right) \tag{4.3}$$

$$H = k_3\left(\frac{\lambda}{2\pi r}\right)^3 + k_4\left(\frac{\lambda}{2\pi r}\right) + k_5\left(\frac{\lambda}{2\pi r}\right) \tag{4.4}$$

where again λ is the wavelength and r is the distance from the loop. These equations are similar to Equations (4.1) and (4.2) except that the roles of E and H are reversed. The near-field/far-field interface is still where $\lambda = 2\pi r$ and the ratio of E to H in the far field is still constant. In the near field, H dominates and the field in this region is called an induction field.

4.4 WAVE IMPEDANCE

The ratio of E to H in a propagating field is called the wave impedance. The term *impedance* applies to sinusoids and suggests a general phase relationship. For a radiating field this expression is not totally appropriate because E and H are in phase. In a radiating field, E has units of volts per meter and H has units of amperes per meter. The ratio E/H obviously has units of ohms, does not depend on dimension, and the vectors are everywhere perpendicular to each other, so the wave impedance could be described in terms of ohms per square, where the square has any dimension.

The ratio of E to H in free space for a plane wave is 377 ohms. For the loop radiator this ratio reduces as the distance from the radiating source is reduced. For example, at 10 MHz the near-field/far-field interface distance is 4.7 m. At a distance of 2.85 m the wave impedance is 188 ohms. If the loop were replaced by a dipole, the wave impedance at this same distance would be increased by a factor of 2, or 754 ohms.

The term *wave impedance* does not imply that the impedance can be directly measured in space. It has meaning when a transmission line must match impedances to drive a radiating antenna. It also has meaning when a shield must reflect wave energy. A low-impedance field is much more difficult to shield than a high-impedance field. Shielding is discussed later.

4.5   EFFECTIVE RADIATED POWER

Assume that a power P is radiated uniformly in all directions from a radiating antenna. The surface area at a distance of R m is $4\pi R^2$. Poynting's vector summed over the surface of the sphere must equal the power P. In equation form,

$$P = (E \times H)\,4\pi R^2 \qquad (4.5)$$

In free space $E/H = 377$ ohms. Substituting H in Equation (4.5) and solving for E yields

$$E = \frac{1}{R}\sqrt{30P} \qquad (4.6)$$

   In most transmission systems it is expensive and unnecessary to send energy in all directions. Antenna systems are designed to avoid transmission toward the sky, out to sea, or into a mountain. Directivity is obviously related to application. In radar the radiation is a narrow beam looking for a target. When radiation is received, it is impossible to determine the nature of this directivity. The only important thing is the field strength at the point of reception. This field strength can be equated to an equivalent source radiating uniformly in all directions. This power level is called the effective radiation power (ERP).

   Radar provides a good example of ERP. Assume the E field from a radar pulse is 10 V/m at a distance of 10,000 m. This requires an ERP of 0.33 GW. If the beam occupies 1% of the total spherical surface, the actual radiating power is only 3.3 MW. If the power is pulsed for 1 ms every second, the average power is only 3.3 kW. The interference in a circuit is that of a 0.33-GW source. The radar power bill is for 3.3-kW load.

4.6   SHIELDING OF WAVE ENERGY

A plane electromagnetic wave is reflected from a large conducting surface in the same way a wave is reflected at the end of a transmission line. If the direction of travel is perpendicular to the conducting surface, the E field must be tangential to the conducting surface. The tangential E field on a perfect conductor must be zero since it would represent an infinite surface current. The conducting surface thus appears as a short circuit. To set the E field to zero at the surface and to return the incoming energy, a second wave is superposed on the first. This wave travels in the reverse direction. When the conducting surface is thin, as in a painted or electroplated surface, some of the wave energy can propagate through the conductor.

A measure of the shielding provided by a specific conductor geometry is called shielding effectivity (SE). The ratio of field strength taken before and after the conductors are in place is a measure of SE. The measure is usually expressed in dB.

$$SE = 20 \log \frac{E_1}{E_2} \tag{4.7}$$

where E is the electric field strength. If SE = 60 dB, it means that the field is attenuated by a factor of 1000 when the shield is in place.

4.7 OHMS PER SQUARE OR SURFACE IMPEDANCE

The potential drop across a square of conducting material with a uniform current flow is a constant for a square of any size. To see that this is true, consider the resistance of a conductor of any cross section. The resistance is the resistivity ρ times the length divided by the cross-sectional area:

$$R = \frac{\rho l}{A} \tag{4.8}$$

Consider the square in Figure 4.1. If l is the square size, then $A = lt$, where t is the material thickness. Using this value for A in Equation (4.8) yields

The Concept of Ohms Per Square

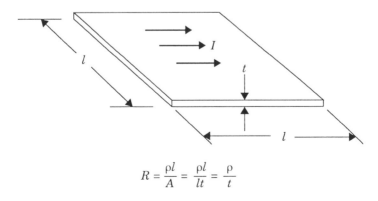

$$R = \frac{\rho l}{A} = \frac{\rho l}{lt} = \frac{\rho}{t}$$

R does not depend on dimension l.

Figure 4.1 A square of conducting material.

$$R = \frac{\rho}{t} \qquad (4.9)$$

which is independent of the square size. Ohms per square is often written as $\Omega/\square$.

The ohms per square for copper and iron for various thicknesses is given in Table 4.1. These values are very low even for frequencies above 10 MHz. This fact makes a large conducting surface a very important tool in interference control. A current of 1000 A flowing uniformly across a sheet of copper only 1.0 mm thick creates a potential drop of only 0.369 V at 1 MHz. At power frequencies the potential drop is only 17.2 mV. For this reason a conducting surface can correctly be called an equipotential surface. Others prefer to call this surface a ground plane. A 1.0-mm-thick copper sheet is very fragile. For practical reasons conducting planes are usually made considerably more robust.

The low impedances in Table 4.1 imply a uniform current flow. When currents concentrate at a connection, the area of the connection no longer provides a low impedance. An increase in current density means that there is a large magnetic field. This implies an inductance storing energy, and in a circuit sense there will be a voltage drop proportional to frequency and current level.

4.8   REFLECTION AND WAVE IMPEDANCE

Every experimenter knows how difficult it is to shield against low-frequency magnetic fields. The reason relates to the fact that at power frequencies the fields are definitely near induction fields. The wave impedance is very low because the interface distance is some 500 miles away. The reflection mechanisms in a transmission line parallel the reflection of field energy at a conductive surface. If the terminating impedance is a big mismatch, the reflection is

Table 4.1   Ohms per Square for Copper and Iron

Frequency	Copper			Steel		
	$t =$ 0.1 mm	$t =$ 1 mm	$t =$ 10 mm	$t =$ 0.1 mm	$t =$ 1 mm	$t =$ 10 mm
10 Hz	172 $\mu\Omega$	17.2 $\mu\Omega$	1.72 $\mu\Omega$	1.01 mΩ	101 $\mu\Omega$	40.1 $\mu\Omega$
100 Hz	172 $\mu\Omega$	17.2 $\mu\Omega$	3.35 $\mu\Omega$	1.01 mΩ	128 $\mu\Omega$	126 $\mu\Omega$
1 kHz	172 $\mu\Omega$	17.5 $\mu\Omega$	11.6 $\mu\Omega$	1.01 mΩ	403 $\mu\Omega$	400 $\mu\Omega$
10 kHz	172 $\mu\Omega$	33.5 $\mu\Omega$	36.9 $\mu\Omega$	1.28 mΩ	1.26 mΩ	1.26 mΩ
100 kHz	175 $\mu\Omega$	116 $\mu\Omega$	116 $\mu\Omega$	4.03 mΩ	4.00 mΩ	4.00 mΩ
1 MHz	335 $\mu\Omega$	369 $\mu\Omega$	369 $\mu\Omega$	12.6 mΩ	12.6 mΩ	12.6 mΩ
10 MHz	1.16 mΩ	1.16 mΩ	1.16 mΩ	40.0 mΩ	40.0 mΩ	40.0 mΩ

guaranteed. If there is a match in impedances, energy is not reflected. A low-impedance wave and a low-impedance surface are somewhat matched, and the energy is not easily reflected. It takes a very thick piece of steel to attenuate the magnetic field significantly near a power transformer.

4.9 SHIELDING EFFECTIVITY

When an electromagnetic wave strikes a conductive surface of finite thickness, energy is reflected and absorbed. The energy that enters the conductor can re-reflect many times and exit the material at each reflection. For most situations this re-reflection process is secondary. The key terms in reducing the penetration of the wave are the reflection loss R and absorption loss A. Shielding effectivity is simply

$$SE_{dB} = A_{dB} + R_{dB} \qquad (4.10)$$

The reflection process is similar to the mismatch found in transmission line theory. The loss in dB is

$$R_{dB} = 20 \, \text{Log} \frac{Z_W}{4Z_B} \qquad (4.11)$$

where Z_W is the wave impedance in ohms (more accurately, ohms per square) and Z_B is the surface impedance in ohms per square. When the wave impedance is low, the reflection loss is obviously smaller. Equation (4.11) assumes an infinite conducting surface, so it can only be used as a guide to what happens in a practical problem.

For example, if the wave impedance is 377 ohms and the surface impedance is 2 ohms per square, the reflection term contributes 33.5 dB to the shielding effectivity. If the field has a wave impedance that is a factor of 10 lower, or 37.7 ohms, the reflection term is reduced by a factor of 10 (20 dB) to 13.5 dB.

4.10 ABSORPTION LOSS AND SKIN DEPTH

An electromagnetic wave that enters a conductive plane is attenuated exponentially with penetration. The changing B field that is present causes an E field in the conductor. This E field generates current that can dissipate energy. This current causes a magnetic field that is dependent on the permeability of the material. The attenuation in the conductor thus depends on frequency, conductivity, permeability, and dielectric constant. For most conductors the dielectric constant can be considered unity. The field intensity at a depth h is given by an equation of the form

$$A = e^{-\gamma h} = e^{-\alpha h} e^{j\omega t} \tag{4.12}$$

where γ consists of real and complex parts. The real part represents attenuation, and the complex part represents the sinusoid.

The attenuation expressed in dB is

$$A = 20 \operatorname{Log} e^{-\alpha h} = -\alpha h \operatorname{Log}_{10} e = -8.68 \alpha h \text{ dB} \tag{4.13}$$

The constant α is equal to

$$\alpha = \sqrt{\pi f \mu \sigma} \tag{4.14}$$

When $h = 1/\alpha$, the attenuation factor is 8.68 dB. This depth is called one skin depth. At two skin depths the attenuation is 17.36 dB.

The constants for copper in Equation (4.14) are:

$$\mu = 4\pi \times 10^{-7} \text{ H/m}$$

$$\rho = \frac{1}{\sigma} = 1.724 \, \mu\text{ohm-cm}$$

$$\sigma = 0.580 \times 10^8 \text{ A/V·m}$$

Note that henries = volt·seconds/ampere. With these constants and dimensions, the attenuation constant $\alpha = 0.117$/mm at 60 Hz.

Copper has a skin depth of 8.55 mm at 60 Hz and 0.066 mm at 1 MHz. Skin depth at another frequency can be calculated by multiplying by the square root of the frequency ratio. Skin depth given by Equation (4.12) is for a plane wave reflecting from an infinite conducting plane. The skin depth equations for other conductor geometries can be found in the literature. Equation (4.14) is adequate for most considerations.

The total shielding effectivity SE is the sum of reflection loss and attenuation loss when these terms are expressed in dB. If the reflection loss is 50 dB and the attenuation loss is 30 dB, the SE is 80 dB. This means that the field intensity after the shield is in place is reduced by a factor of 10,000. Shielding effectivity charts are available in the literature for copper and steel for various thicknesses. Near-field SE is shown for radiating sources positioned 3 and 30 m away. These distances are selected to agree with published testing requirements.

Skin depth in iron is a function of permeability. In earlier discussions, permeability had several definitions because the hysteresis loops in iron are highly nonlinear. At the highest application frequencies, permeability tends to fall off rapidly. This occurs even for ferrite cores. Obviously, if permeability is changing, then skin depth is also changing. Published curves of SE are often theoretical and assume an unchanging permeability, which does not exist.

Beware of charts where the curves are for materials having a permeability of 10,000 at 1 GHz.

If the SE calculation is 10 dB or less, there may be a need to add a correction term *B*, which allows for multiple reflections. This low an SE number probably means that the shielding is already inadequate. This *B* factor can be found in the literature.[*]

The measure of SE for various materials is not a simple task. First, an infinite testing surface is not available; second, a meaningful radiating source is difficult to arrange. When the test is made with a standard setup, the results give a good idea of expected performance. Shielding to reflect and absorb radiated energy is most often used in small geometries, and any SE measurements are at best only a relative indicator of performance.

A question often asked involves the grounding of this shield. SE as discussed here does not require grounding. Circuit performance within a shielded enclosure may require an ohmic connection for different reasons. Ferrite materials are insulators since the magnetic domains are held in a ceramic binding material. In this case grounding has no meaning.

4.11 APERTURES

Apertures abound in electronics. There are ventilation holes, seams, panel lamps, displays, CRTs, connectors, and so forth. An electromagnetic field can penetrate these apertures, making any shielding relatively ineffective. For most sheet metal enclosures the total SE is dominated by apertures, not by the material itself. For thin materials the conducting skin can be considered one more aperture.

The simplest assumption one can make about an aperture is that it will allow full-wave penetration when a half-wavelength equals the maximum dimension of the aperture. For example, at 100 MHz a half-wavelength is 1.5 m. In a steel building the distance between beams is greater than 1.5 m, and the signal from an FM radio station at 98 MHz can easily penetrate the building.

When the half-wavelength of a radiated signal is longer than the maximum opening, the attenuation is simply the ratio of half-wavelength to maximum dimension. If a radio station broadcasts at 1 MHz, the half-wavelength is 150 m. If the maximum diagonal distance in the beam structure is 5 m, the attenuation ratio is 5/150, or one part in 30. This means that the radio signal is attenuated by a factor of 30. This phenomena can be easily observed when you drive into an underground parking structure or tunnel. Unless an antenna system has been installed, AM radio stations will fade and FM signals will not.

---

[*] See D. White, *A Handbook of Electromagnetic Shielding Materials and Performance*, ICT, Gainsville, VA.

Electronic equipment that must operate in high-field environments or that require very low levels of interference may require treatment of every aperture. Digital circuits often radiate at frequencies above 100 MHz. If an aperture is to attenuate this signal by a factor of 1000, the aperture must be smaller than 0.15 cm. Circuits may have to operate reliably in a radar beam or near a radio transmitter. In these cases aperture shielding is important.

The infinite conducting plane and a single aperture pose a very difficult mathematical problem. Assume a plane wave is directed at an aperture with an *E* field parallel to the long dimension of the aperture. The field pattern near the corners of the aperture will have a lot of fine structure. The penetrating wave is no longer a plane wave. The field strength on the far side of the aperture is certainly not constant, and this is the ideal case. Now complicate the problem by assuming that the wave does not arrive perpendicular to the conducting plane and that the *E* field is at an angle with respect to the aperture orientation. In the real world there is no infinite plane, and the electronics is a box with other walls that also have apertures and circuits are mounted inside. This is truly impossible to handle mathematically. Any attempt to measure the SE by placing an antenna inside the box will change the very field that is to be measured.

The best way to handle a problem of this complexity is to consider a worst-case scenario. If this analysis shows that the aperture treatment is adequate, the problem is solved. This approach is used by most designers, although the assumptions made are often criticized as oversimplifications. The assumptions are:

1. The wave direction and polarization are always optimum.
2. The penetrating wave is a plane wave and is not attenuated by the presence of circuits.
3. Re-reflections from the box walls do not take place.
4. Any field coupling to the circuit is optimum.
5. Fields arriving from different apertures add directly.

In other words, it can never be worse than this. Assumption 1 is not that critical. If the wave polarization is off by 45°, the coupling is only reduced by 3 dB, or 30%. This level of error is rather incidental when the entire problem is considered.

4.12 MULTIPLE APERTURES

Widely separated apertures allow wave energy to independently enter an enclosed area. The simplest assumption is that these fields add. If a seam attenuates a field by 40 dB (1%) and a ventilation hole attenuates the same field by 30 dB (3%), the fields that enter sum to 4% of the original field. To convert this to dB, simply take 20 times the Log of 0.04, or −28 dB. This summation should not allow the internal field to exceed the level of the external field since there can be no gain. The measure is usually made with the *E* field, although the *H* field is sometimes considered. The assumption is that the attenuated wave is a plane wave and the ratio of *E* to *H* is still 377 ohms, which allows the *H* field to be simply evaluated.

Many apertures are close together or form an array. When this occurs, the apertures are said to be dependent. In this situation the array acts as one aperture. The criteria for independence involves the circulation of surface current. If current can freely circulate around each opening, then these openings are independent. A seam closed by a series of evenly spaced screws is considered a dependent set of apertures. A group of ventilation holes is also dependent and functions as one aperture.

A very narrow aperture has additional attenuation, but additional SE is difficult to quantize. The safest assumption is the worst-case scenario, which ignores the small dimension.

A wire mesh can be used to close a large aperture. The individual mesh squares form dependent apertures since current cannot flow freely around each cell. The net result is that the screen looks like one small aperture. There is a catch, however. To be effective the wire screen must bond around the entire perimeter of the opening. If bonding is partial, another aperture is formed. Bonding is often unfamiliar to circuit designers. Merely grounding the screen is not sufficient. The screen must make a good electrical connection on the entire perimeter or it will not function as an attenuator. Bonding may require a conductive gasket and hardware that maintains the contact under pressure. The conductive surface cannot be painted or anodized, and it should not oxidize over time. An installation that looks good optically may not perform at all.

Not all wire meshes are qualified to close an aperture. Each wire intersection must make an ohmic connection or the grid will not function. Chicken wire is acceptable, because each intersection is tin dipped. Aluminum screen is suspect because the wire can easily oxidize.

Wire mesh is usually not suitable for closing off a CRT opening. The writing pattern and the grid pattern form a moire pattern which can be very disturbing to the viewer. Conducting glass and very thin conductive depositions can be used to provide some SE. The problem is a balance between optical transparency and SE. Another method is to use optics to view the CRT and allow the optical path to provide waveguide attenuation. This type of attenuation is discussed in the next section.

4.13 WAVEGUIDES

Electromagnetic energy can be transferred down a coaxial path at all frequencies. At a high enough frequency, energy can propagate inside a hollow conducting cylinder without a center conductor. The energy is said to be transported by waveguide action. The lowest frequency of transport occurs when the half-wavelength is approximately equal to the width of the waveguide. Below this frequency, energy that enters is attenuated exponentially with depth. A 5-cm waveguide has its lowest mode of transport at about 3GHz. For frequencies below 3 GHz the wave energy is simply attenuated.

The attenuation in dB for a waveguide that is operating below its lowest frequency is

$$A_{dB} = 30 \frac{l}{d} \tag{4.15}$$

where d is waveguide width and l is waveguide depth. The waveguide is said to be operating beyond cutoff, or WGBC. A waveguide can be very effective in attenuating unwanted field energy without closing the opening. If $l/d = 3$, the attenuation is 90 dB, or a factor of 30,000. If a conductor is threaded through the waveguide, it becomes coax and no longer serves as an attenuator.

Honeycomb construction makes an effective ventilation system. The l/d ratio can be greater than 10, providing attenuation of over 300 dB per cell. The honeycomb cells must be well bonded to each other, and the perimeter must be well bonded to the aperture opening for the honeycomb to be effective. Hardware to effect this bonding is usually supplied by the manufacturer.

Waveguide attenuation can be provided at an aperture by adding depth. As long as the aspect ratio l/d is maintained, the opening can provide significant attenuation. This approach is used to treat the aperture formed around the door of a screen room. Finger stock makes many contacts, and the seam has a depth of several inches. If a channel is provided, the finger stock can be protected and the path length for waveguide attenuation can be made long. In the design of plastic cases that are eventually electroplated, the seams can be made quite deep to provide additional SE.

The honeycomb provides a set of independent apertures, since current can circulate freely around each cell. The field that enters each cell is given by the ratio of half-wavelength to cell opening. If the cell diameter is 1 cm and the half-wavelength is 100 cm, the field is attenuated by a factor of 100, or 40 dB. If the aspect ratio l/d is 5, the WGBC attenuation is 150 dB. The total field attenuation per cell is thus 190 dB. If there are 50 cells, each one can couple field energy. This reduces the attenuation by 34 dB. The final SE of the honeycomb is thus 156 dB, or an attenuation factor of 6.3×10^7. Waveguide attenuation is thus a powerful tool.

4.14   A THOUGHT EXPERIMENT

Consider a large metal pan with a lid. If a small battery-operated FM receiver is placed in the pan, the radio will play when the lid is open but will stop when the lid is in place. Now bring a 3-ft insulated wire into the pan. The radio now plays with the lid closed. The lesson is simple. If there is one open aperture or if one lead is brought into the enclosure, there is a field inside of the enclosure.

To keep unwanted fields out of an enclosure, all leads carrying field energy must be filtered. To carry the experiment one step further, connect the wire to the pan but leave a loop of wire in the pan. The radio will still play with the lid in place. If the lead is grounded to the outside of the pan, the radio will not work. The location of the connection and the size of the loop affect the field in the pan.

Hence, a filter must be located at the entry point of the conductors to effectively stop the entry of field energy. If it is inside the pan, the loop formed by input leads radiates energy into the pan going around the filter. This is a good example of where a circuit diagram fails to show what can happen. If the pan is a ground, a circuit diagram simply shows a connection. It makes a great deal of difference if the connection is outside the pan or inside, and, if inside, the size of the loop area formed is also important.

4.15 PASSIVE FILTERS

The previous section discussed the location of a filter to reflect or absorb any high-frequency energy being carried on leads. The reference conductor for the filter must be the bulkhead at the point of entry. A remote ground reference point for the filter will limit its performance. Passive filters connect shunt capacitors to the metal case of the filter, which is the reference conductor for the filter. Any lead connecting the case of the filter to the circuit represents an inductance in series with these capacitors. Any series inductance significantly alters the filter characteristics. The assumption here is that published data can often include frequencies up to 300 MHz.

It is interesting to see how a filter manufacturer might test a filter. A reference ground plane is used. The oscillator, meters, and filter are carefully bonded to the ground plane. Shielded leads are used for all interconnections. It is easy to see that the ground lead to the filter is essentially zero in length. A filter will not meet published specifications unless this same bonding is provided.

Filters for power line applications are sometimes provided with snap-in hardware, which may not provide an adequate bond from the filter case to the mounting surface. If the surface is painted or anodized, the connection might also be inadequate. If a filter is truly needed, it must be properly bonded or it cannot meet all of its published specifications. This is another example of where a connection that tests zero ohms with an ohmmeter is not a bond. Testing a bond for microohms and nanohenries of inductance is not simple, yet these are the parameters that must be considered.

Connectors that have filters on each lead are available, but they are expensive and the filtering action, because of size constraints, is minimal. They are used in military equipment that must pass very severe environmental tests.

From this discussion one could easily conclude that power line filters are not always needed. Most of them are improperly installed, yet equipment seems to function. There is truth to this statement, but filters will continue to be a part of most designs because it is the proper thing to do. Fortunately, there is a second line of defense that involves circuit design and layout. Even if fields are present, many things can be done to avoid coupling. We discuss them in later chapters.

4.16 FILTERING POWER AND EQUIPMENT GROUND CONDUCTORS

Power line filters must treat all power conductors before the filtering can be effective. Nature pays no attention to our color codes, and all conductor pairs can carry energy. The NEC does not allow filters in the equipment ground path because they can limit fault current. Thus, the equipment ground must terminate on the outside of the equipment. If it is brought inside, interference can radiate from the lead into nearby circuits. Here is another example of where a circuit diagram is inadequate. The location of this connection is the key to keeping radiation out of the equipment. Figure 4.2 shows how a filter should be mounted so that it is electrically external to the equipment. The equipment ground terminates outside the equipment enclosure.

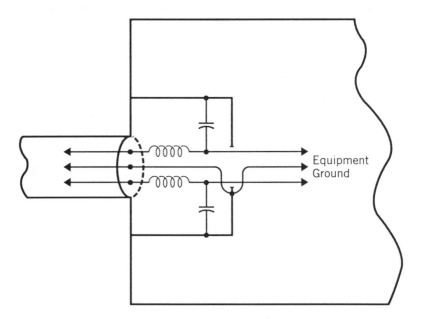

Power Line Filter — Correct
"Green Wire" Treatment

Figure 4.2 A proper power filter installation.

Interference can often be generated inside of equipment. It may be necessary to filter this interference so that it is not transported out of the equipment onto the power conductors. Without proper filtering this energy can flow between power conductors or between conductors and the conduit enclosing them. Breaking the conduit connection in any way is unsafe and not allowed by the NEC. Power transformers provide a degree of interference isolation, but even a shielded transformer has a primary-to-secondary leakage capacitance of, say, 5 pF. This capacitance represents a reactance of 3200 ohms at 10 MHz. In a typical unshielded transformer a primary-to-secondary capacitance of 1000 pF is not uncommon. This is a reactance of 16 ohms at 10 MHz, a very low impedance. A switching regulator can generate content at this frequency and cause currents to flow back into the power conductors.

Often it is desirable to limit interference that can flow back into the power conductors. Any filter used here must terminate at the bulkhead, but on the *inside* surface. If properly designed, a filter can serve to block the flow of energy in both directions. L-type filters have an inductor as their first element. This form of filter is preferred over a capacitor input filter. If the interference involves a low-impedance source, an input capacitor can draw a large current. Large currents can radiate, and most bulkheads are not perfect ground planes.

4.17 FIELD COUPLING INTO CIRCUITS

When a plane wave propagates parallel to the plane of a circuit loop, coupling is optimal when the *H* field enters the loop and the *E* field is in the plane of the loop. The coupling can be calculated with either field, and the answer should be the same. The easiest calculation involves the *E* field because voltages are obvious. The *H* field calculation involves Faraday's law and the rate at which the *H* field is changing.

A typical loop is shown in Figure 4.3. Coupling is always assumed to be optimal since this is a part of the worst-case analysis approach used in design. Even if the direction of the *E* field is at an angle of 45°, the result will be off by only 30%.

The *E* field is conservative, which means that the field intensity times every increment of distance around any closed loop must sum to zero. The *E* field in the direction of propagation is zero, which means that the only distances that contribute to the summation are at the ends of the loop. The *E* field intensity differs at the ends because the wave is traveling at a finite velocity. In this summation the *E* field times distance is simply volts. When the summation is made, the result is positive when the direction of travel is in the direction of the field. The greater the spacing between conductors in the loop, the larger is this voltage.

The voltage at the end of the loop at any instant in time depends on the part of the wave at that end of the loop. In general, the longer the loop the larger the voltage difference will be. The maximum voltage difference occurs when the loop length is one half-wavelength. At this length the peak voltage coupled to the loop is twice the peak value observed at one end.

The coupled voltage is thus proportional to conductor spacing and loop length. In other words, it is proportional to loop area. Where does this voltage appear? It appears across the points of highest impedance in the loop. If the loop is used to send a signal to a circuit, the field adds voltage directly to this signal.

As frequency increases, the half-wavelength becomes less than the loop length and the coupled voltage begins to fall off. At a full wavelength, the coupling is zero. It is easy to show that the coupling is a maximum at odd multiples of a half-wavelength and zero at even multiples. For a worst-case analysis this canceling effect is disregarded. At frequencies where the half-wavelength is less than the loop length, the coupling is assumed to be a maximum.

Assume a field strength of 10 V/m. The loop in question is 10 cm by 100 cm. What is the coupling at 10 MHz? The half-wavelength is 1500 cm. The voltage calculated along the 10-cm dimension is 1 V. The maximum coupling is twice this value, or 2 V. The ratio of loop length to half-wavelength is 15:1. This means that the coupled voltage is $\frac{1}{15}$ of 2 V or 0.13 V.

4.18   A PUZZLING MEASUREMENT

A metal aircraft is positioned 100 ft from a radiating antenna. The frequency of the radiator is 100 MHz. A voltmeter is used to trace lines of equal potential on the skin of the aircraft. One lead of the voltmeter is connected to a sheet of metal under the aircraft. The voltmeter indicates voltages that range from 0 to 60 V. At first glance this seems plausible.

Now ask an embarrassing question. If the skin impedance in ohms per square is roughly 1 milliohm, how can there be 60 V of potential difference on the skin? If this were true, the surface currents would have to be thousands of amperes, and this is not about to happen from a 10-W transmitter.

The answer is simple. There is no significant voltage gradient on the skin of the aircraft. The voltmeter circuit is the culprit. The leads are simply coupling to fields near the aircraft. If the voltmeter leads were positioned so that there was no loop area, the voltages measured on the skin of the aircraft might not exceed 1 mV. Using the metal under the aircraft as a reference conductor is not valid.

4.19   EMC REGULATIONS

Electromagnetic compatibility (EMC) regulations and standards exist for most major countries. As the technology grows, so does the list of regulations. A design engineer should first focus on these standards so that the design does

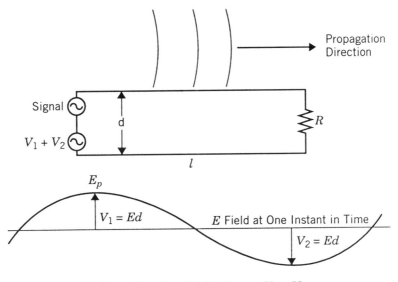

Maximum Coupling Total Voltage = $V_1 + V_2$.
When l is a half wavelength.

Figure 4.3   Field coupling to a loop.

not have to be redone later. In many companies the standards are set internally so that the market conditions are automatically met.

In the United States, regulations are set for commercial equipment by the FCC. Its primary role is to control conducted and radiated interference so that users of radios and televisions are not disturbed and equipment is basically safe. Equipment sold for military applications must meet military standards that are constantly updated. Some of these standards are used as guidelines to establish good practice.

In Europe the standards are slowly changing to accommodate the emerging common market. The German body that regulates EMC is the Verband Deutscher Elektrotechniker (VDE). The British regulating body is the British Standards Institute (BSI). Some 44 nations belong to an international group known as CISPR. This acronym is French and stands for International Special Committee on Radio Interference. The IEC is a central European regulations organization of which CISPR is a part. CISPR has begun to issue Euro-norms to govern the electronics sold in the European common market. Products manufactured in the United States will be required to meet these Euro-norms. In general, these requirements are tighter than the FCC regulations.

Testing falls into four categories: conducted emission (CE), conducted susceptibility (CS), radiated emission (RE), and radiated susceptibility (RS). Not all tests are applicable to every product, and different testing levels are required for different categories of equipment. To radiate, some conduction is required, so the tests are not necessarily independent. The specifications simply state how the tests are to be made. Often the tests are on individual

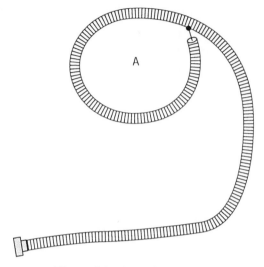

Figure 4.4   A "Sniffer" cable.

pieces of equipment, and no tests are defined for assembled hardware. Testing defines levels, bandwidths, measuring distances, and configuration.

4.20   THE "SNIFFER" AND OTHER DEVICES

A designer needs a few tools to measure radiation without resorting to costly testing in a laboratory. When the equipment goes for certification, the designer should be fairly certain there will be no problems. The tools need not be calibrated or certified, and a simple calculation will usually relate signal pickup to field strength. Some of these simple tools are commercially available.

One simple tool is the "Sniffer." It is a loop formed at the end of a coaxial cable. The cable is simply used as the input to an oscilloscope. The cable shielding excludes E field coupling. The H field that is sensed is proportional to the loop area. The voltage measured is proportional to the rate of change of flux. The loop is simply moved around the equipment to check for leakage flux. Radiation should be checked near apertures, near connectors, and along cable runs. The "Sniffer" geometry is shown in Figure 4.4.

Simple loops of wire can also be used as a measuring tool. The number of turns and the diameter of the loop relate to the frequency and level of the signals observed. If power-related phenomena are involved, the loop may involve up to 100 turns of fine wire.

Dipole antennas are relatively easy to build. A whip antenna mounted on a metal plate can be moved around to detect radiation. If necessary a small battery-operated amplifier can be placed on the plate to handle small signals. The gain can be shaped so that the response is flat over a portion of the spectrum.

CHAPTER 5

CALCULATING TOOLS

5.1 WAVEFORMS AND FREQUENCY

The sine wave is the only waveform that can be processed through a general linear circuit and come out unchanged. All other waveforms are modified. The term *frequency* can mean a sinusoid or a repetition rate. In field theory and circuit theory it usually implies a sinusoid.

A repetitive waveform can always be expressed as a series of sinusoids. This group of sinusoids is called a Fourier series. It is interesting to sum the following set of sinusoids:

$$\sin(x), \frac{\sin(3x)}{3}, \frac{\sin(5x)}{5}, \frac{\sin(7x)}{7}, \ldots \tag{5.1}$$

Figure 5.1 shows the sum as each term is included.

The infinite series adds up to a square wave. If every term in Equation (5.1) is changed to a cosine function, the sum is still a square wave but the wave is now symmetrical about the midpoint of the positive step. Every term in Equation (5.1) is called a harmonic, and the first term is called the fundamental.

When a square-wave voltage is the input signal to a circuit, one analysis is to treat each harmonic separately. The results at each frequency are then added to provide a composite output signal. This approach is viable in any linear system. It is not always a simple task to reconstruct the repetitive waveform that results.

The x in Equation (5.1) is a phase angle. For a sinusoidal signal the phase angle changes with time. At the end of one cycle the phase must rotate through

Harmonics That Make up a Square Wave

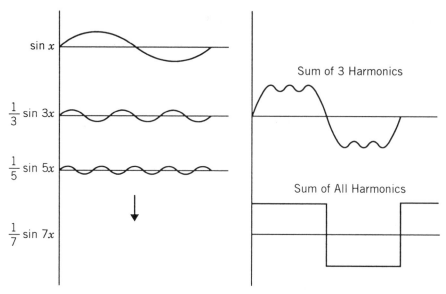

Figure 5.1   The components of a square wave.

360° or 2π radians. If x is replaced by $2\pi ft$, then, when time $t = 1/f$, $x = 2\pi$, or one full cycle. When $t = 2/f$, then $x = 4\pi$, or two full cycles. As t increases, the sinusoid progresses through all angles. The period of the fundamental is $1/f$.

5.2   A REALISTIC SQUARE WAVE

The square wave in Figure 5.1 was ideal as it transitioned between values in zero time. In every practical situation this time is finite. Rise time is usually specified as the time between the 10% and 90% points.

The Fourier series for this repetitive waveform is more complex than the simple square wave. If the wave crosses the horizontal axis at $t = 0$, as in Figure 5.1, the series has no cosine terms. If the rise and fall times are different, then some cosine terms are present. In a practical situation the shortest transition time determines the impact on a circuit. For an analysis the shortest time is selected, and it becomes the new rise and fall times. This is in keeping with the idea of doing a worst-case analysis.

For a perfect square wave the amplitudes of all harmonics decrease proportionally to frequency. The amplitude of the third harmonic is one-third the amplitude of the fundamental. The amplitude of the seventh harmonic is one-seventh the amplitude of the fundamental, and so forth. This inverse relationship between frequency and harmonic amplitude is often plotted logarithmically.

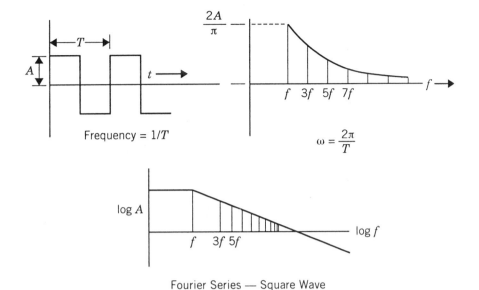

Fourier Series — Square Wave

Figure 5.2 A logarithmic plot of harmonic amplitudes for an ideal square wave.

Figure 5.2 shows a logarithmic plot of harmonic amplitude versus frequency for an ideal square wave. Note that the harmonic amplitudes have been reduced by a factor of $2/\pi$ so that the harmonics produce a square wave with a unity amplitude. Also note that in this logarithmic plot the points representing the harmonic amplitudes all lie on a straight line.

The intersection of this sloping line and unity amplitude is called the first corner frequency. The curve in Figure 5.2 is called the harmonic envelope. For the ideal square wave, all of the harmonic amplitudes fall on this line. The presence of an envelope does not imply content at all frequencies. Note that the first corner frequency is not the fundamental frequency.

When the square wave has finite rise and fall times, the harmonic amplitude plot is shown in Figure 5.3. At some of the harmonics the amplitude may be zero. The exact solution shows that for the lower harmonics, the amplitudes

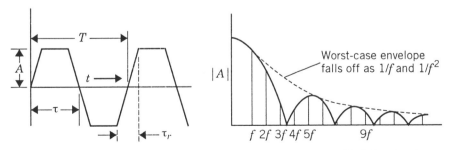

Figure 5.3 A plot of harmonic amplitudes for a square wave with finite rise time.

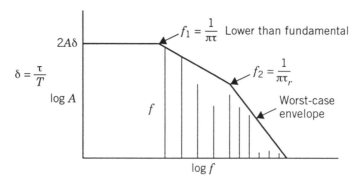

Figure 5.4   A harmonic amplitude envelope for a square wave with finite rise time.

are always less than the envelope in Figure 5.2. Above a frequency $1/\pi\tau_r$ the solutions show that the harmonic amplitudes generally fall off as the square of frequency. When data are plotted logarithmically, a straight-line envelope can be constructed that guarantees containment of all harmonic amplitudes. This plot, in Figure 5.4, shows a first corner frequency at $1/\pi\tau$, as in Figure 5.2, and a second corner frequency at $1/\pi\tau_r$. This second corner frequency is a key number in treating many interference-type problems.

5.3   A TRAIN OF SHORT PULSES

A repetitive series of short pulses has a harmonic spectrum as shown in Figure 5.5. The straight-line segment of the envelope has an amplitude of $2A\delta$, where δ is the duty cycle, given by τ/T, and A is the amplitude of the pulse train. The first corner frequency is now higher in frequency than the fundamental and many of the lower harmonics.

It is interesting to consider the Fourier series of a pulse train when δ gets very small. Each pulse is, in effect, isolated in time. The harmonic amplitudes that make up this pulse train are reduced in amplitude, and the number of harmonics below the first corner frequency is increased.

5.4   THE SINGLE EVENT

Many electrical events are pulselike, including lightning, electrostatic discharges, switch closures, and arcing. These single events all consist of a frequency spectrum. If the results of the previous section are taken to the limit, then it is no longer possible to talk about harmonics. There is signal at every frequency, including dc, and the amplitude at any frequency is zero. The

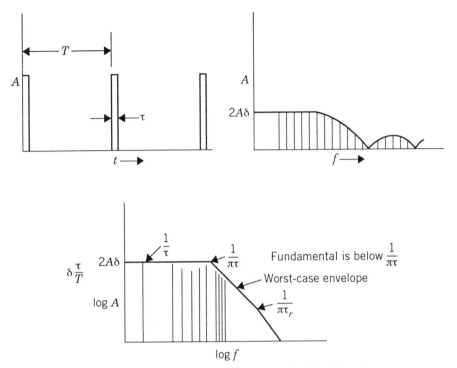

Figure 5.5 A series of short pulses and its Fourier series.

amplitude is finite only for a defined bandwidth. In a spectrum analyzer, for example, the results are often given for a narrow bandwidth. The measurement is usually given in terms such as volts per kilohertz.

A pulse entering a circuit must be treated by considering the response at every frequency and integrating the results. This technique can be handled by many computer programs. Later a simple worst-case approach is described that approximates the magnitude response of a circuit to a single pulse.

The envelope used in Section 5.2 can still be applied to single events. The assumption is made again that the rise and fall times are equal to the shortest of the two times. The envelope has first and second corner frequencies as before, but the straight-line portion has a different amplitude unit. The Fourier spectrum of a single pulse lasting a time τ with a rise time τ_r is shown in Figure 5.6. Note that the amplitude $2A\tau$ can be read as amplitude $\times$ time (volt-seconds) or as volts/hertz.

5.5 PULSE OR LEADING-EDGE COUPLING

The response of a circuit to an impulse requires a complex calculation. In many cases an engineer would like to get a feeling for this coupling without

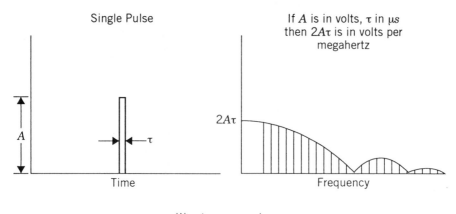

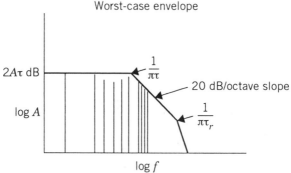

Figure 5.6   The Fourier spectrum of single pulse.

a complex analysis. The response to a pulse is usually nonlinear, and a detailed sinusoidal analysis makes little sense.

Most coupling mechanisms involve a loop area and a changing flux. This occurs when a plane wave couples to a signal loop or when a magnetic field couples to a loop of wire. In both situations the coupling is proportional to frequency. The spectral envelope for an impulse falls off proportionally to frequency from the first corner frequency to the second corner frequency (see Figure 5.6). The increase in coupling with frequency and the fall-off in amplitude with frequency cancel each other, with the result that a pulse couples uniformly to most circuits at all frequencies out to the second corner frequency. This fact can be used to approximate the coupling into a loop. One assumes a single-frequency analysis at the second corner frequency. The sine wave root-mean-square (rms) amplitude is assumed to be the peak value of the pulse. It might be argued that the peak of the pulse should correspond to the peak of the sinusoid. The additional factor of $\sqrt{2}$ adds a bit of safety to the calculation.

This same approach can be used for repetitive waveforms where there is a fast leading edge. The rise time yields a second corner frequency which can be used for a calculation. For example, a digital clock signal has a rise time of 10 ns. The second corner frequency is 31.8 MHz. If a circuit loop carries a 50-mA clock signal, this amplitude and frequency can be used to estimate radiation or loop coupling.

5.6 NOISE AND PULSE MEASUREMENTS

The units on the vertical scale in Figure 5.6 need to be explained further. First assume the amplitude of the pulse is 1 V. If the pulse width is 1 ms, the vertical scale can be read as volts per kilohertz. When the pulse width is 0.5 ms, the vertical scale is then read as 1 V per 2 kHz. A subtle question arises as to whether this voltage can be read as 0.5 V/kHz.

When two unrelated (noncoherent) sinusoids are summed on an rms basis, the result is given by the square root of the sum of the squares. If one voltage is 3 V at 1.313 kHz and the second unrelated voltage is 4 V at 1.773 kHz, the sum is 5 V, not 7 V. Consider a square wave made up of odd harmonics. These voltages are coherent, because they must be timed correctly to make up the square wave. The sinusoids that go together to make up a single pulse are also coherent even though they are everywhere dense.

Consider a spectrum of pure noise. Here the sinusoids that make up the spectrum are everywhere dense but not related. In this case the signals do not add linearly. For example, if the noise in a 2-kHz band is 1 V, the noise in a 1-kHz band is not 0.5 V but 0.707 V. Now the question posed earlier can be answered. The vertical scale for a pulse can be read 1 V per 2 kHz or 0.5 V/kHz, and both are correct. This is true because the sinusoids that make up the pulse are coherent.

The vertical scale can represent volts per kilohertz, amperes per megahertz, E field per hertz, and so forth. The units used are properly a point measure. This scale in no way implies that a specific bandwidth must be used. When a pulse interferes with a narrowband receiver, this narrow bandwidth must be considered. The signal in any bandwidth is the integral of amplitude frequency density over the frequency range of interest. When the spectrum is reasonably flat, the product of bandwidth and spectral density provides a good approximation.

The wideband noise specified for an integrated circuit is given in terms of voltage per square root hertz. A typical number might be $4\,\text{nV}/\sqrt{\text{Hz}}$. This noise is random and thus noncoherent. To calculate the noise in a megahertz bandwidth, simply multiply the noise figure by the square root of the bandwidth. In this case $4\,\text{nV}/\sqrt{\text{Hz}} \times 1000$ shows that the noise in 1-MHz bandwidth is 4 μV.

In practical applications the amplitude response of any device falls off with frequency above some upper frequency. It is common practice to consider the -3-dB point as the bandwidth. In a typical application the response might fall off linearly above this frequency. To be accurate, a noise calculation must then give consideration to the shape of the response curve. When a noise figure is given, the way the measurement was made should be stated. If the measure is made with an RC filter having a given -3-dB point, this should be stated. The noise contribution above the -3-dB point can be as much as 15%.

The noise of an integrated circuit is complicated by a phenomena known as $1/f$ noise or shot noise. In applications where signals above a few hertz are filtered out, the noise is no longer proportional to the square root of frequency. As the frequency decreases, the noise increases. This phenomenon is critical in dc amplifier applications such as weighing machines and strain-gauge work. This type of noise is described in IC manufacturers specifications.

If there is a peak in the amplitude response of an amplifier, the noise contribution is emphasized in that portion of the spectrum. Noise calculations involve the sum of the squared voltages. It is easy to see that a large peak in the response can add significantly to the noise figure. The trained eye can easily detect this type of noise.

5.7   THE DECIBEL

As common as the decibel is, it is often misused, abused, and misunderstood. The electromagnetic community uses the dB extensively. Specifications, regulations, and performance data are all given in dB. The dB is a convenient logarithmic scale that allows very large and very small values to be displayed on the same graph.

The bel is the base 10 logarithm of a power ratio. In telephony the decibel (dB), or 10 times the logarithm of power ratio, was invented. This was a convenient unit because a 1-dB change in power sound level is just discernible to most listeners. The ratio must always use terms with the same units to be dimensionless.

Most signals are measured in volts or amperes, not watts. Power can always be calculated from volts or amperes if the impedance level is known. Equation (5.2) shows how voltage ratios are used to calculate dB:

$$\text{dB} = 10 \, \text{Log} \frac{P_1}{P_2} = 10 \, \text{Log} \frac{V_1 R_2}{V_2 R_1}$$
$$= 20 \, \text{Log} \frac{V_1}{V_2} + 10 \, \text{Log} \frac{R_2}{R_1} \tag{5.2}$$

If the impedances are equal, the dB measure of voltage ratio is

$$\text{VdB} = 20 \, \text{Log} \frac{V_1}{V_2} \tag{5.3}$$

It is common practice to measure a parameter in terms such as dB volts or dB amperes. The measure involves a reference value in the denominator of the implied ratio. For example, 100 V is 40 dB volts. The ratio implied is 100 V/1 V: 100 V = 40 dB volts. If a parameter is expressed in dB, a subscript is often used, such as V_{dB}. Another notation is dBV.

5.8 THE dB EXPANDED

A logarithmic scale is almost a necessity in engineering, and the dB scale is the one of choice. The gain of operational amplifiers is given in dB. The closed-loop gain might be 60 dB and the open-loop gain 130 dB. In both cases impedance is not a consideration. Strictly speaking, this use is incorrect. Attempts have been made to remedy this abuse by using a different notation, but the efforts have failed. It is up to the reader to know how to interpret the meaning.

Some equations include product terms of resistance, distance, frequency, and so forth. To handle these terms it is easier to use a logarithmic base so that the calculation involves sums and differences rather than products and divisions. The result is expressions such as dB ohms, dB meters, and dB hertz. A 40-dB ohm resistor is just a 100-ohm resistor. A 0-dB resistor is a 1-ohm resistor. For example, 2 A flowing in 5 ohms is 6 dB amperes and 14 dB ohms. IR is simply $6 + 14$ dB volts, or 10 V.

The dB notation can be confusing. The letter m usually means milliwatt. The notation dBm always means dB milliwatts. In this case the reference value is 1 mW, and the calculation involves 10 Log $P/1$ mW, where P must be in milliwatts.

5.9 REFERENCE dB UNITS

A ratio must always be formed before a dB value can be determined. The denominator term is the reference value, and all units must be defined in the notation. For voltage the reference value is usually obvious, say 1 V or 1 mV or 1 μV. In the case of gain, units may not be required. When bandwidth and field strength are involved, the units can get rather complex. The reference might be 1 μV/m/kHz or 1 mA/m/MHz. This last term is read milliamperes per meter per megahertz.

5.10 THE dB AND Vu SCALES

It is accepted practice in audio engineering to use a logarithmic voltage scale to measure sound level. Sound is measured in volume units (Vu), where one Vu unit is a decibel. In the early days of audio engineering, all signals were processed through balanced transformers with controlled impedance levels. A measure of sound voltage level was indeed a measure of sound power. Present designs use low-impedance sources and high-impedance terminations, and power transfer is not the issue. Voltage levels as monitored on a Vu meter are used to measure sound level.

The 0-dB reference point on many ac voltmeters is 1 mW in 600 ohms. This is 0.774 V rms. This level was also at one time the zero reference for Vu meters. The pointer on a well-designed Vu meter has a defined ballistics response (natural frequency and damping factor). A repeatable dynamic character is very useful when voice and music signals with expected form factors are monitored. As audio engineering evolved, standard voltage levels also changed. In most commercial equipment the 0 Vu reference level has been set at +4 dBm or 1.23 V rms. In broadcast equipment the level has been set at +8 dBm or 1.94 V rms. The zero point on the Vu meters is set by varying the resistance in series with the meter. Changing this resistor changes the dynamic character of the meter.

The reference levels used in handling audio signals are related to "head room." Voice and music by its very nature have peak values of signal that often exceed five times the peaks of the rms value. At 1.23 V rms signal peaks can exceed 8.7 V. This dynamic range is obviously exceeded if 5-V power supplies are used.

5.11 FEEDBACK BASICS

Feedback techniques can be used as a design tool to raise input impedance, lower output impedance, reject common-mode signals, improve frequency response, improve gain accuracy, and reduce distortion, among other things. The basic amplifying tool in design is the integrated circuit (IC), which takes the form of an operational amplifier (op amps). Op amps, available in hundreds of versions, have positive and negative input terminals and a great deal of forward gain—often exceeding 1 million at low frequencies, and this excess gain provides feedback to tailor the application and improve performance.

An easy way to understand the feedback process is to consider the circuit in Figure 5.7. The positive input terminal is connected to circuit common. The negative input terminal is connected to the output through resistor R_2 and to the input through resistor R_1. If the gain of the amplifier is 1 million and the full-scale output signal is 10 V, the maximum signal on the negative input terminal is 10 μV. For all practical purposes this point is held fixed at 0 V and cannot change voltage. This point is often called a virtual ground.

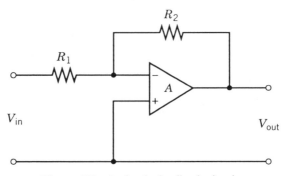

Figure 5.7   A simple feedback circuit.

When an input signal V_{IN} is applied to input resistor R_1, the input current can be calculated since the voltage at input 1 is near zero. By Ohm's law the current is V_{IN}/R_1. This current cannot flow into the amplifier because there is no voltage at this point. The only path for this current is in R_2. The voltage at the output is simply the current times R_2, or $V_O = -V_{IN}(R_2/R_1)$. The ratio of V_O/V_{IN} is the gain, and in this example it is close to the ratio of R_2/R_1. If R_2 is 20,000 ohms and R_1 is 5000 ohms, the gain is -4.

The gain of the circuit in Figure 5.7 is exactly

$$G = \frac{-R_2}{R_1 + \dfrac{R_2}{A} + \dfrac{R_1}{A}} \qquad (5.4)$$

For $A = 10^4$ the gain is -3.9980. If $A = 10^5$, the gain is -3.99980. Hence, the larger the value of A, the closer is the gain to -4.000.

The gain in excess of 4 is called the feedback factor. The feedback factor determines the gain accuracy. If A is 10,000, then the gain in excess of 4 is 2500. The gain error is simply one part in 2500, or 0.04%. The gain of 4 reduced by 0.04% is 3.99980, which agrees with the result of using Equation (5.4). Again, the gain in excess of the required gain determines the gain error. This assumes of course that the resistors are accurate.

5.12 FEEDBACK—A MORE FORMAL LOOK

Feedback, as the word implies, is the return of output signal to the input. It is convenient to use the symbology shown in Figure 5.8. Here a fraction of the output signal is summed into the input. In general, the signal that is fed back must subtract from the input signal. If feedback reenforces the input signal, the result is usually an instability. This fraction of fed-back signal is denoted by β. The summation process is perfectly general and does not depend on any circuit technique. The negative input terminal is called a summing

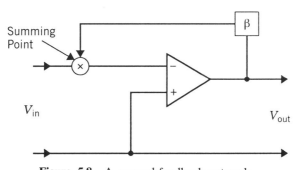

Figure 5.8   A general feedback network.

point because it responds to the sum of the input signal and a fraction of the output signal. The gain of this circuit is

$$G = \frac{-A}{1 + A\beta} \tag{5.5}$$

where $-A$ is the open-loop gain of the amplifier. It is easy to see that when A is large the gain approaches $-1/\beta$. This gain is called the closed-loop gain. The closed-loop gain can be fully controlled by resistor values and not by the gain of the amplifier.

5.13   FEEDBACK AND INTERNAL ERRORS

The circuit making up the gain in Figure 5.8 may consist of several stages. Figure 5.9 shows two gain segments with an internal summing point. This new summing point can be used to show how internal error signals are attenuated by feedback. The gain to normal signals is simply

$$G = \frac{-A_1 A_2}{1 + A_1 A_2 \beta} \tag{5.6}$$

The gain to an internal error signal is

$$G_1 = \frac{1}{A_1 \beta} \tag{5.7}$$

This equation shows that the gain to the error signal is the normal gain $1/\beta$ divided by A_1. This means that internal errors are reduced by the gain preceding the point of injection.

To illustrate this point, consider an internal stage with a 1% linearity figure. If the gain preceding this point is 100, the error at the output is reduced by a factor of 100, or 0.01%. Errors produced or sensed at the input are not

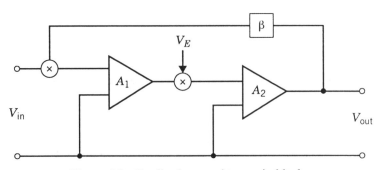

Figure 5.9   Feedback around two gain blocks.

reduced by any preceding gain. These errors are simply amplified by the closed-loop gain $1/\beta$.

Internal errors can include dc drift, noise, hum coupling nonlinearities, and phase shift. Feedback reduces these errors by preemphasis. An error signal representing the negative of the anomaly is added to the normal signal at the input summing point. The error signal is amplified and used to cancel the anomaly. The signals inside of a feedback system are distorted ahead, not after, of an anomaly. Inside a feedback system the point of overload or distortion occurs ahead of the point of difficulty. During troubleshooting, the circuit in overload is being told to try and correct a downstream problem.

5.14 FEEDBACK AND CIRCUIT STABILITY

All amplification is limited by the ability of circuits to supply reactive energy. In a typical integrated circuit some parasitic capacitance exists across every transistor collector resistor. This capacitance reduces the gain of each stage at high frequencies. It is very important that overall gain as a function of frequency be carefully controlled. If this is not done, the circuit will oscillate when feedback is applied.

Each capacitance that adds to the attenuation of signal in an amplifier adds phase shift. In the limit one capacitance adds attenuation that is proportional to frequency and a phase shift of 90°. When several capacitances contribute to the attenuation, the phase shift increases even more rapidly with frequency. A phase shift of 180° represents a reversal of sign for the gain.

Equation (5.5) shows a very interesting property of every feedback system. The denominator term becomes zero when $A\beta = -1$, which means the gain goes to infinity and the amplifier is unstable. In most cases $A\beta$ may never be exactly -1. It is a good rule of thumb that if $A\beta > 1$ when the phase shift reaches 180° the feedback system will be unstable. The requirement for absolute stability is called the Nyquist criteria, which is fully discussed in textbooks.

The worst case occurs when β is unity or all of the output is fed back. This condition places the most severe restraints on the amplifier amplitude response and associated phase shift out to a frequency where the gain is unity. IC amplifiers with this class of stability are said to be stabilized or unconditionally stable. Users must be careful not to place an excessive capacitive load on the amplifier output because this adds to the phase shift and can result in instability, particularly for small values of closed-loop gain.

Controlled total phase shift implies a controlled amplitude response. There is no way to avoid excessive phase shift if the amplitude response falls off too rapidly with frequency. A phase shift of 90° corresponds to an attenuation rate of 20 dB per decade. One way to achieve a stable response is to allow one capacitance to dominate the amplitude response. The problem of stability and gain rolloff is important for both designer and user.

5.15   GAIN BANDWIDTH PRODUCT

The stability issue involves the open-loop gain of an amplifier at all frequencies. If an IC amplifier has a gain-bandwidth product of 10^6, it means that at 1 MHz the open-loop gain is unity. Extrapolating backwards, this means that the amplifier has an open-loop gain of 10 at 100 kHz and an open-loop gain of 100 at 10 kHz. Extending this concept further, the open-loop gain is 1 million but only at frequencies below 1 Hz, a dismaying fact if there is an expectation of high open-loop gain at, say, 10 kHz. For a closed-loop gain of 100 and a feedback factor of 10 at 10 kHz, the gain-bandwidth product must be 10 MHz. This assumes an IC amplifier stabilized for unity gain.

5.16   COMMON-MODE REJECTION AND DIFFERENTIAL AMPLIFIERS

A powerful tool in electronic design is the differential amplifier. This type of circuit can reject common-mode signals, which represent a large class of interference. Common-mode rejection (CMR) is a measure of how well this interference is rejected. Unfortunately, CMR is not as simple as one specification at a single frequency.

A common-mode signal is often the average voltage on a group of conductors with respect to some reference conductor. If a group of conductors carries signal to an amplifier and a field couples signal to the signal-earth loop, this coupled voltage is a common-mode signal. The reference conductor can be the output common of the amplifier. A lightning pulse carried on a group of power conductors is a common-mode signal with respect to the earth. Consider the case where signals are transported between two pieces of grounded equipment. If the common of the receiving equipment is the common-mode reference conductor, the common-mode signal is the ground potential difference.

Not all common-mode signals are interference. If the input signal is 10 V and one side of the signal is common, the average signal on the two leads is 5 V. This is technically a common-mode signal, but it surely is not interference. The power conductors can carry 0 and 120 V. The average value with respect

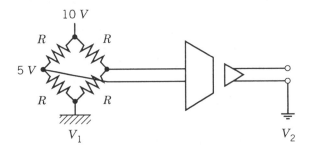

Figure 5.10   Two common-mode signals.

to equipment ground is 60 V. This is the common-mode voltage, and it must not be rejected. This signal is also not interference.

Two common-mode signals might be present simultaneously. An example might be a bonded strain-gauge amplifier operating between two grounds. One-half of the bridge excitation is a common-mode signal, and the ground potential difference is the other. These two common-mode signals have different reference conductors. This situation is shown in Figure 5.10.

5.17   THE DIFFERENTIAL AMPLIFIER—A GENERAL DISCUSSION

The A and B inputs of an oscilloscope are differential inputs when the gains are set equal but opposite in sign. When the same signal is applied to the two inputs, the output is zero. In this sense the oscilloscope rejects the common-mode component. If a normal signal could be monitored with these same inputs, the input and output signals would share the same common conductor. This class of differential signal processing cannot reject the common-mode signals discussed in the previous section.

The most general differential amplifier must process signals in two electrostatic environments. The input leads are connected to an input circuit that has a reference conductor. The output leads are grounded to a second piece of equipment, or at least connected to an oscilloscope for observation. In general, the two reference conductors cannot be shorted together to eliminate any common-mode signal present. This point was carefully considered in Section 2.16. The electronics in the input is enclosed (shielded) by the input common, and the electronics of the output is enclosed (shielded) by the output common. The shielding in each section should control the flow of circulating current from the individual power transformers so that current is directed away from signal conductors. This general case is shown in Figure 5.11. The obvious problem is how to couple signals between the two electrostatic regions. Whatever the nature of the solution, if it rejects common-mode signals it is a differential amplifier. The difference is amplified, and the common-mode component is not amplified.

The bridge between the two circuits can involve transformers, optical devices, or high-impedance circuits. The transformer approach has been used for years by audio and telephone engineers. Optical coupling can be used, although it is difficult to maintain linearity. The high-impedance approach is preferred because it is the most economical.

In Figure 5.11 the source impedance is assumed to be unbalanced by a resistance R. Any interference current flowing in this resistance couples unwanted signal directly into the first amplifier. If this current flows as the result of a common-mode signal, this coupling defines the CMR ratio.

Common-mode voltages can cause current to flow in the unbalanced input resistance. What impedance is required to meet a reasonable common-mode

The Fundamental Instrumentation Problem

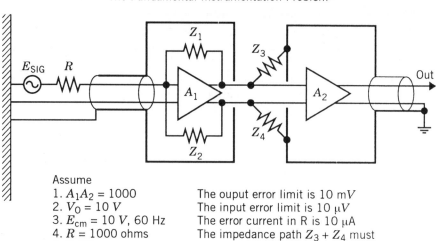

Assume
1. $A_1A_2 = 1000$ The ouput error limit is 10 mV
2. $V_O = 10$ V   The input error limit is 10 μV
3. $E_{cm} = 10$ V, 60 Hz   The error current in R is 10 μA
4. $R = 1000$ ohms   The impedance path $Z_3 + Z_4$ must
5. 0.1% error   exceed 1000 megohms.
                         This is 2 pF at 60 Hz

Figure 5.11   The two-signal environment.

specification? An interesting problem occurs in instrumentation where the gains are often 1000. In Figure 5.11 assume $R = 1000$ ohms, the common-mode signal is 10 V at 60 Hz, the output signal level is 10 V, and the permitted error is 0.1%. A 0.1% error is 10 mV at the output and 10 μV at the input. The current in the 1000-ohm resistor is limited to 10 nA. The common-mode impedance must therefore be 1000 megohms. At 60 Hz this is the reactance of only 2 pF.

The general solution to this isolation problem is too expensive. In most applications the entire input circuit and its power transformer can be eliminated, leaving an input shield and the electronics in the output environment. This solution is shown in Figure 5.12. Note that the transformer coils can still circulate power current in the output common leads. Any voltage developed in these leads is not amplified. Therefore it is possible to build the equipment without multiple shields in the power transformer.

The 1000-megohm common-mode problem has now been transferred to the input impedance of the amplifier. This high impedance is not provided to allow operation from high-impedance sources but to provide common-mode rejection. The input shield must be present and connected to the signal where the signal grounds. The arguments in Section 3.2 still apply. Leakage capacitance out of the input shield must be held to under 2 pF or the CMR figure will not be met. This means that the entire input signal path must be properly handled. Just buying an instrument with a specification does not solve the problem, since it is a systems issue.

This CMR problem can be described in terms of a balanced audio signal. Consider the following parameters: source unbalance, 10 ohms; common-

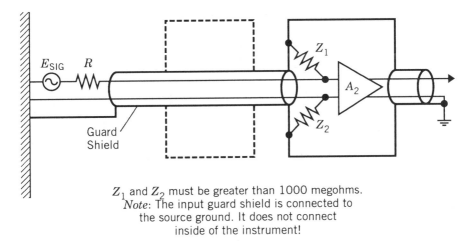

Z_1 and Z_2 must be greater than 1000 megohms.
Note: The input guard shield is connected to
the source ground. It does not connect
inside of the instrument!

Figure 5.12 A practical differential amplifier.

mode level, 1 V at 10 kHz; gain, 10; full-scale signal, 1 V; error limit, 0.1%. The maximum input signal error allowed is 100 μV. The current in the unbalanced impedance is limited to 10 μA. This is a CM impedance of 100,000 ohms. A 10-ohm unbalance at 10 kHz is an inductance of 1.59 μH. A 100,000 ohm reactance at 10 kHz is provided by a capacitance of 15.9 pF. Cable balance and amplifier output impedance balance are part of a system specification that must be considered if the common-mode specification is to be met.

The input shield in Figure 5.12 is usually called a guard shield. It must be grounded where the input signal leads are grounded. See Section 3.2. If the input leads float, the shield must still be connected to the signal. A floating or undefined shield can cause a lot of difficulty that may be hard to detect. A balanced signal midpoint (Wagner ground) is sometimes used. It can be as simple as a resistive divider across the two signal leads. This point can be used for the input shield termination or for grounding the input signal.

5.18 COMMON-MODE REJECTION CONSIDERATIONS

Common-mode signals are not restricted to 60 Hz. They may take the form of power harmonics, pulses caused by load switching, noise from switching power supplies, radio transmitters, and so forth. The behavior of circuits at frequencies other than 60 Hz is thus an important topic. Circuits separated by large distances are most likely to be affected by high-frequency interference.

Common-mode signals that can be handled by electronic circuitry are limited in amplitude. Simple amplifier circuits that operate between two 15-V power supplies cannot process common-mode signals plus normal-mode signals that exceed 10 V peak. It is practical to handle a larger dynamic range using more sophisticated circuitry. Available techniques include driven power supplies, transformer-coupled modulators/demodulators, and common-mode attenuators.

Common-mode electronics has all the characteristics of any amplifier. It has a highest frequency where it will develop a full-scale signal, and it has small-signal bandwidth. Common-mode performance is rarely specified as a function of frequency, but it is required if a signal is to be processed without interference.

Common-mode clipping can occur in some applications. After overload the signal should recover to the same value as if the clipping had not occurred. If the common-mode electronics is integral with the normal signal electronics, the recovery time may be extended.

Common-mode performance varies with the gain setting. The CMRR should be highest when the gain requirements are highest. Designs can place the CMR process at the output, in an intermediate attenuator, or at the input. In each approach the CMRR varies with gain, depending on how much gain precedes the point of CM rejection. CMRR is usually referred to the input (rti) of an amplifier, where it can be compared with an input signal. A gain 1000 amplifier might meet a CMRR of 120 dB at 60 Hz rti with an input line unbalance of 1000 ohms. This means that a CM signal of 10 V adds 10 μV to the input signal. If the gain before the rejection point is reduced by 10, the CMRR rti is only 100 dB. If the gain is reduced after the rejection point, the CMRR rti remains unchanged.

A total CMRR specification involves frequency, amplitude, maximum full-scale frequency, gain settings, and line unbalance in any combination. As stated earlier, a single number does not describe the performance.

5.19 A SIMPLE DIFFERENTIAL AMPLIFIER

The differential amplifier is a general tool that takes on many different forms. The general case discussed involves fairly sophisticated circuitry. These designs are needed in general instrumentation applications. The simplest feedback structure that rejects common-mode signals is shown in Figure 5.13. This

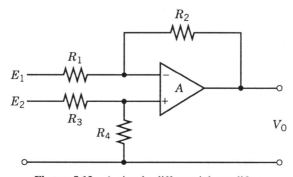

Figure 5.13 A simple differential amplifier.

circuit can provide gain to normal-mode signals and no gain to common-mode signals. The only condition placed on the circuit is that $R_1/R_2 = R_3/R_4$. If E_2 is zero, the gain to E_1 is $-R_2/R_1$. If E_1 is zero, the gain to E_2 is R_2/R_1. If $E_1 = E_2$, the amplifier sees a pure common-mode signal and the output is zero.

5.20 FORWARD REFERENCING AMPLIFIERS

When the simple differential amplifier in Figure 5.13 uses four equal feedback resistors, the normal gain is unity. This circuit is shown in Figure 5.14. The signal source can be balanced or unbalanced, and only the difference signal is amplified. When the signal source is single-ended, one side of the signal is a common or ground. This common lead can be connected to either input of the differential amplifier. If the inverting input is connected to this common, the differential amplifier is noninverting. This amplifier rejects any ground difference of potential. In effect, it references the signal to a new forward ground. Ground potential differences are often power related and are the source of hum and general power "hash."

The source impedance of the input signal must be considered in maintaining the ratio of R_2 to R_1 and R_4 to R_3. The higher the four resistors are in value, the less sensitivity there is to this problem. Typically, 10,000-ohm metal film resistors can provide a CMRR of 10,000 to 1 from dc to several kilohertz, assuming the source impedance is less than 1 ohm. A potentiometer can be added to the circuit to set the CM balance. If both gain and CMR are to be adjusted, two potentiometers are required.

Forward referencing amplifiers can be used to isolate video signals with frequency response up to 10 MHz. The application might involve an interconnection between separately powered pieces of equipment. If the integrated circuit has sufficient bandwidth, the matching resistors should be in the 1000-ohm range.

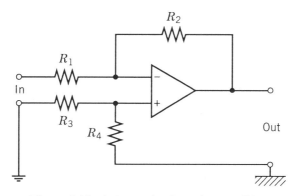

Figure 5.14 A forward referencing amplifier.

In accurate A/D converter designs, small ground potential differences can cause errors. A forward referencing amplifier can be added to the circuit if it is not integral to the A/D converter design. This approach eliminates errors caused by power current flow or inadvertent digital feedback.

The power supply for this form of isolating amplifier must be referenced to the output common. If a group of signals needs to be isolated, then one bipolar power supply can be shared. Typical applications might involve tape recorders, analog inputs to a computer, audio mixing, and so forth.

CHAPTER 6

INSTRUMENTATION AMPLIFIERS

6.1 INTRODUCTION

Amplifiers designed to condition signals from strain gauges, thermocouples, resistance-temperature detector (RTDs), positional sensors, pressure gauges, and so forth are traditionally called instrumentation amplifiers. They are also called universal signal conditioning amplifiers because they are used for general-purpose testing. They usually have gains from 1 to 10,000, frequency response from dc to 100 kHz, operation with 300 V of common-mode signal present, selectable low-pass filters, noise rti below 2 μV, and provisions for calibration, monitoring, and signal substitution. They may be supplied with an internal excitation source and reference voltage.

The circuit techniques used to build these accurate instruments find application in many other areas. It is for this reason that a chapter is devoted to this class of equipment. The feedback techniques, the component selection, the layouts, the testing, and the systems considerations are all vital topics. These techniques are used in audio engineering, industrial control, transducer design, system design, and measurement.

Instrumentation amplifiers are generally dc amplifiers, which adds to the design difficulty. On the other hand, it is often easier and less expensive to design a dc system with one coupling capacitor than to design an ac system with its associated problems of overload recovery. Fewer capacitors mean smaller size and reduced cost. In some designs the dc gain can be zero, which means that drift is no issue.

Basic system problems always exist when signals are sensed in one region, carried by cable, and displayed or stored in a second region. These problems exist even if the signal is digitized within the transducer. The issues are accuracy, calibration, transport, bandwidth, and economics. These issues remain regardless of how the technology is applied. If the production volume is high enough in any product area, the technology can serve to reduce price and improve performance. Care must be taken so that each new approach considers the basics and improves the end results.

Many early instrumentation systems consisted of carrier amplifiers where a strain-gauge bridge was excited at ac. This allowed the use of transformers to couple signals and provide isolation, to provide conversion from balanced to single-ended amplifiers, and to realize very low noise rti. These systems suffered from inaccuracies caused by carrier cross-coupling and demodulation errors. The maximum bandwidth was around 1 kHz. When dc amplifiers made their debut, they were often chopper stabilized, with the entire instrument devoted to gain. These vacuum-tube designs were single-ended with obvious limitations. With the transistor revolution it became possible to design smaller instruments, make them differential, and add a few conditioning features. With the advent of integrated circuits, the amplifying portion of the design became a small part of each instrument. Designs incorporated many important aspects of signal conditioning, including excitation, filters, A/D converters, calibration, balance, computer control, monitoring, signal substitution, and so forth.

The market for general-purpose instrumentation is relatively small. Because of the volatile nature of this business, it has been the domain of small companies. A small group of engineers, familiar with a complex set of issues, has played the specification game and supplied these instruments to industry. This art is not easily found in the literature because the engineers are busy making changes and fighting the odds of keeping a business afloat.

The problems of signal conditioning are only one part of the larger systems issue. Users rarely question passive transducers or simple cable. They also trust the recording devices which are supplied by major manufacturers. This leaves many of the subtle issues to be resolved in the instruments. If there are problems, the finger of blame has no place else to point. Unfortunately many of the issues that can cause trouble are not brought out in the specifications. This happens because the engineer writing specifications combines material available from several manufacturing brochures and probably has never designed equipment. The manufacturers are often aware of many issues they do not wish to discuss in their description and specifications. Meeting more and more restrictions is not the way to stay alive in business. Also, mentioning a specification may force the competition to respond in an improper way. Further, one manufacturer may grasp issues not recognized or understood by another. Getting manufacturers and specifiers together is not easy, although many attempts have been made.

Instrumentation-type IC amplifiers exist. They must be imbedded in additional circuitry before they are useful, and require, for example, power supplies, gain-changing structures, input protection, input bias adjustments, connectors, and output buffering. By the time these circuits are added, the IC cost is incidental. These IC devices are often lacking in bandwidth, gain range, signal levels, common-mode levels, or slew rate. In specific applications these parameters might not be critical. It is also easier to use a ready-made design than to learn how to build a successful input stage.

6.2 THE TESTING DILEMMA

Instruments delivered to a facility for acceptance are usually evaluated singly. Evaluating groups of instruments in the actual system can be very difficult and expensive and may only be done after the system is operational. This cannot be done unless a turnkey system is purchased. The evaluation of individual instruments is usually done by a calibration laboratory equipped to measure static parameters. Some measurements are simple, but the test equipment must be more accurate than the test. Many specifications require measurement equipment with accuracies of 0.01%. At this level of accuracy many subtle issues exist, which are often poorly understood by the testing facility as well as the manufacturer.

One of the most difficult philosophical aspects in testing and in specification writing involves the interrelationship between many parameters. For example, gain may be specified with an accuracy figure of 0.1%. The manufacturer knows there will be measurement errors, so it tries to keep the gain accuracy error below this value, say 0.08%. A tighter design adds significant cost. The gain test often involves unstated standard conditions, such as nominal line voltage and temperature. This point is usually understood, but it is often not stated in the specifications. Temperature coefficients and line voltage stability are specified but treated as separate issues.

Gain in a dynamic instrument is just that, a gain at dc and ac. At what frequency is gain still accurate to 0.1%? This question is complex because gain measurements at ac require different measuring techniques. Obtaining sine wave sources with distortions below 0.03% and voltmeters with 0.03% accuracy may be difficult. For this reason it is standard practice to measure gain at dc and completely ignore the ac issue. To some extent this is acceptable, but there are designs where the 10-Hz response can be off by 1% and the dc and 100-Hz response can be correct.

Unfortunately, measuring gain at dc is also not easy. When a known voltage is placed at the input terminals, the gain is not simply the ratio of output to input voltages. Here are a few of the difficulties. The amplifier has noise, drift, and linearity errors. Each of these errors adds signal to the output measurement. To understand the problem of measuring gain is to understand a great deal about systems error and parameter interrelation. There is a technique for quickly measuring gain, but it does not involve a sine wave generator, ac or dc voltmeter, or a precision dc voltage source. It involves a square-wave generator and some precision resistors, and this requires a carefully constructed test fixture. Then there are subtle questions such as: does the gain change with common-mode level? If so, what is the common-mode frequency?

Gain itself is usually measured without an output load and without an input line unbalance. Further, in a differential amplifier there is gain from either input terminal or both terminals. Also, the line unbalance can occur on either input, or the source impedance can be balanced and not zero. Measuring gain for a range of line source impedances implies taking a lot of data. Measuring gain without adding input line unbalances may not properly evaluate the instrument.

Gain and linearity are usually measured with respect to a full-scale signal. When the expected signal is one-half of full-scale, it is only correct to assume that the errors are all doubled with respect to this half-scale. This is a good argument for requiring small gain errors, since signals often fall well below full scale. In testing, the answer is unknown, and for this reason there must be plenty of "head room" to ensure getting results with some accuracy.

Gain is only one parameter, yet it is a complicated specification. A single number is only a guideline. It should be used with care and measured with understanding. Gain of an instrument is only one aspect of system gain. Every device that processes the signal has an equal set of complications. It may be next to impossible to determine full end-to-end gain accuracy in a system because this must involve the transducer. A meaningful figure on accuracy or accuracy probability is possible, but obviously difficult to obtain.

Gain accuracy in some systems is not an issue. If the gain and zero errors are determined in calibration just before a run, the impact of temperature, line voltage, loading, cable length, or drift are all taken into consideration. This reduces the cost of instrumentation and places the accuracy burden on the system design. For audio engineering the accuracy requirements may be a few percent, and most of the issues described here do not exist. A lot has to do with the nature of the application. Meeting very difficult specifications in preparation for every possible application may be impractical, unnecessary, and expensive.

In most system applications where computers can be used in calibration, stability, not accuracy, is the key issue. Corrections can be added later or just before the data are recorded. Repeatable performance is a direct indication of stability. Having both repeatable and accurate performance is a luxury that adds to the cost. When possible, system accuracy rather than component accuracy should always be the purpose of calibration. Then factors such as drift, temperature, line voltage, source unbalance, and loading can be calibrated out. It is possible to have each segment of a system self-calibrate, but errors resulting from interconnection are not accounted for. Of course, if system calibration is not available, then the stability of each segment is the only choice. If the data cannot be corrected efficiently, then stability and accuracy are both necessary. At this point calibration and certification measurements are critical and not easily made.

6.3 THE HIGH-INPUT-IMPEDANCE INPUT STAGE

In Chapter 5 the general problem of differential amplification was discussed. The nature of the input impedance and the treatment of the input shield were discussed. Leakage capacitances of only 2 pF violate the 1000-megohm input impedance requirements needed to meet an acceptable common-mode specification. Of course, this was for an instrument with a gain of 1000 and with a 1000-ohm line unbalance. For nearly balanced systems where high gain is not required, the problem is simpler. The high input impedance for common-mode rejection is usually provided in the input amplifier. The common-mode rejection process can use a balanced attenuator/differential amplifier located after the initial gain stage.

High input impedance can be obtained from potentiometric feedback. Consider an input transistor followed by gain in an operational amplifier. A fraction of the output signal is fed back to the transistor emitter and the input impedance is raised. This circuit is shown in Figure 6.1. The gain is determined by the attenuator R_1 and R_2, where $\beta = R_2/(R_1 + R_2)$. If the base-emitter impedance is 500 ohms before feedback and if the feedback factor is 1 million, the input impedance is 500 megohms.

The operation of this circuit can be described by noticing that the positive input to the IC amplifier is fixed and defined by the voltage across zener diode D_1. Because of the high gain that follows, the negative input terminal must be very near this same voltage. This means that the transistor collector voltage is known. The voltage drop and, thus, the current in the collector resistor R_C are thus controlled by the power supply voltage. The collector current is

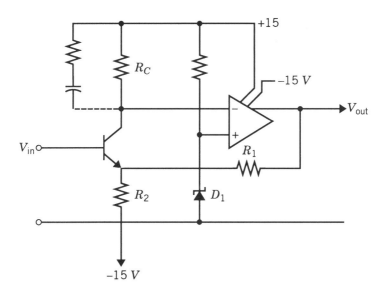

Figure 6.1 Raising the input impedance.

supplied through the feedback resistors from the output of the amplifier. The emitter voltage is defined by the input signal and the transistor operating characteristics. The offset voltage at the output of the IC amplifier is considered later. Assume the feedback fraction β is 0.2. If the input voltage changes by 1 V, the output of the IC amplifier changes by 5 V.

The collector current in the input transistor is held nearly constant for all signal levels, which contributes significantly to system stability. As the input signal changes, the available collector-to-emitter voltage changes. The maximum positive input voltage occurs when there is no collector voltage left for the operation of the input transistor. If the collector voltage is, say, 7 V, the maximum input voltage may be 6 V.

The collector current in this type of circuit can be set to about 0.1 mA. If the transistor beta is 400, the input base current is about 0.25 μA. Most of this current can be supplied from the circuit itself rather than requiring it to flow in the source impedance. If only 1% of this current flows in a 1000-ohm source impedance, the error is 2.5 μV rti, 0.025% of full scale at a gain of 1000.

If the input stage is a field effect transistor (FET), the gate of the transistor requires negligible current, so current compensation circuitry might not be needed. The penalty is greater noise and drift rti. With a transistor input stage the noise can be held to below 1.3 μV rms rti in a 100-kHz bandwidth. With a FET input stage the noise may be increased by a factor of 10.

The circuit in Figure 6.1 will oscillate unless some forward gain shaping is supplied. A series RC circuit paralleling the input collector can reduce the loop gain for frequencies above, say, 10 kHz. If the collector resistor is 20 kohms, then R can be 2 kohms and the gain is reduced by a factor of 10. If the gain reduction starts at 10 kHz, the value of C would be around 800 pF. Shaping needs to be adjusted until the loop is stable for all gains. For a square-wave test the output signal should have less than 20% overshoot with limited undershoot.

6.4 THE BALANCED HIGH-INPUT-IMPEDANCE CIRCUIT

Balanced operation can be provided by using two input circuits, the only difference being that the feedback attenuator no longer goes to output common but connects the two emitters. See Figure 6.2. When one input has zero signal, the second emitter is held fixed, making it a virtual ground. The feedback attenuator for the active input functions in the same way as the circuit in Figure 6.1. This feedback arrangement is typical of a standard differential instrument amplifier. The gain of this circuit is given by

$$G = 1 + \frac{2R_1}{R_2} \tag{6.1}$$

Note that when $R_2 = \infty$ the gain is unity.

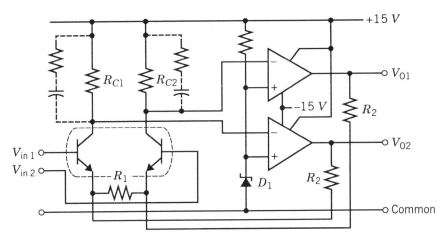

Figure 6.2 A balanced input circuit.

The zero-input-signal reference for the circuit in Figure 6.2 is the common of the power supply. It is preferable to establish a new reference potential defined by the average input signal. This average signal when measured with respect to an output common is the common-mode voltage. This potential can be used to drive the entire input circuit, including power supplies. This average input signal can be found by summing the signals from the two input emitters. The voltage representing this common-mode level is the midpoint between R_3 and R_4. This signal is buffered by a FET follower as shown in Figure 6.3. The power supplies for the input stage and the two operational amplifiers are regulated by zener diodes powered from the driven power supply. Enough current must flow in the zener diode string to supply the input transistor pair and the two operational amplifiers at all signal levels. The zener diodes now move with the common-mode voltage. The FET source follower has an offset voltage that shifts the potential on the zener string. This slight modification to the power supply has no impact on circuit performance.

Driving the power supplies for the input stages with the common-mode signal is a feedback system. There is some gain from the power supplies to the output of the two operational amplifiers. This output signal is fed back to the two emitters which determine the common-mode signal level. The loop gain is not high, but the loop stability must be checked before the design is acceptable.

If transistors T_1 and T_2 are a matched pair built on the same substrate, output voltages V_1 and V_2 will be identical except for a small dc offset. Any small difference can be attributed to transistor pair offset and drift. Matched transistor pairs are available from several semiconductor manufacturers. The feedback structure guarantees that $V_1 - V_2$ is a direct measure of the difference between the two input signals. The fixed offset on these two outputs is a balanced low-impedance common-mode voltage source which can be removed in the next stage.

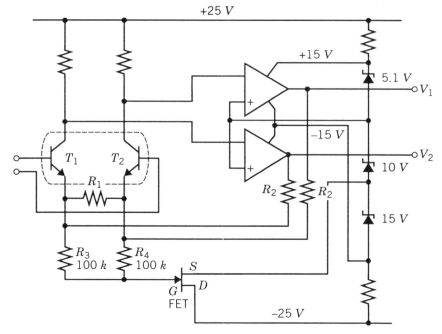

Figure 6.3 A common-mode driver added to the balanced amplifier system.

The power supply providing current to the zener string can use simple series resistors or, preferably, current sources. The load is thereby removed from the driving FET follower, which improves CMR performance. The voltage level (compliance) of this power supply depends on application. If the input common-mode signal with respect to guard shield is 15 V, the input-to-output common-mode voltage is 15 V, the maximum input signal is 10 V, the current source takes 7 V, and the power supplies requires 15 V, then the primary power supply must be greater than the sum of these voltages. It is not uncommon to find ±80-V supplies since this also allows for low-line-voltage operation. The zener power supply string operated from constant-current sources is shown in Figure 6.4.

The driven power supply can now supply bias current to the input transistors. The bias current supplied in this way does not depend on the common-mode level and the input impedance is not compromised. Resistors used to supply this current can be 100-megohm carbon $\frac{1}{4}$ W, available on special order. Note that the common-mode impedance, not the differential input impedance, must be 1000 megohms. In most designs the differential input impedance can be held to 50 megohms at dc. However, this is not an invitation to use high-impedance signal sources, because the input offset current that flows in the source must still be considered.

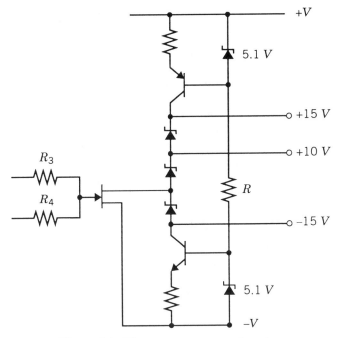

Figure 6.4 The zener power supply string.

6.5 INPUT PROTECTION AND TREATMENT

To avoid instrument overload when the input is left open, high-value resistors are often placed across the input and from input to input guard shield. The guard shield is connected to output common by approximately 1 megohm in case the input is not grounded. These resistors allow a path for bias current to flow. Resistors associated with the input leads are typically 100 megohms. Such high values are used so that input specifications are not compromised. The input leads must be properly guarded (shielded), or overload at high gains is ensured. If a user incorrectly uses only one input lead and floats the other, the instrument might not overload but neither might it perform well.

Input circuits must be protected against both input overload and foldover. Foldover can occur when large input signals overload the input stage. Foldover causes the output signal to appear in range during an overload, and the user may assume that the amplifier is functional. Foldover can be blocked by clamping zener diodes and clamping transistors that detect an excessive input current. Clamping requires a series resistor in each line. Without them the zener clamps could be damaged by a low-impedance source. The series resistors and zeners must be of sufficient wattage to accept prolonged overload. During setup it may take a long time before an overload situation can be corrected, and the instrument must survive this torture. Series resistors must be limited to a few hundred ohms, or their Boltzmann noise will add to the input noise. This noise can be somewhat reduced by bypassing the series resistors with a nominal capacitor.

Input circuits should be protected against high-frequency content coupled to the signal lines from external sources. This protection usually takes the form of small input inductors (10 μH) and small bypass capacitors (200 pF) from each input to guard shield. The components must be small enough so that they do not impact the frequency response of the instrument when operating from some upper limit of source impedance. In most applications a safe upper impedance limit is 10 kohms.

High-frequency energy can enter an instrument on every lead pair, including power, equipment ground, output, monitors, calibration, excitation, and so forth. This energy can be most disruptive if it couples to the input because there is no feedback available to limit its impact. Thus, very small capacitors (20 pF) should be shunted between the base and emitters of the input transistors to force the high-frequency content at this point to be near zero. This technique is far more effective than filtering every lead entering an instrument.

Thermocouple effects can add small dc voltages to the input. Any dissymmetry in the input geometry is a possible source of error. One source of thermocouple pickup occurs when relays are used to switch the input leads. Switching can occur in calibration, signal substitution, or monitoring. Relay and connector contacts should always be gold-plated. Relays should be evaluated for their thermocouple effects before being accepted in a design. Latching relays are preferred because coil heating of input connections is then eliminated. However, latching relays require logic to handle the driving coils in the relay.

Input lead loop area is critical in an instrumentation amplifier. Loop area can couple to both high-frequency fields and power-related magnetic fields. Layouts must give the input loop area first consideration. If the leads go through a relay, that relay must be carefully selected so that its internal construction adds little loop area. When circuit traces are used in the input, spacings of about 12 mils work quite well. Even here leads should never run closer than 2 in. from any power transformer core. When input leads exit through a connector, the pin spacing is critical. The connector shell should be connected to guard shield and no other potential.

The input guard shield should be connected to the input leads where the input grounds (connects) to the source. This connection is the only one that allows the instrument to reject common-mode signals. See Section 5.16. The input cable and any other nearby conductor can support the transport of high-frequency energy. Without filtering, the guard shield brings this energy into the instrument. To limit the impact of this intrusion, the guard shield should be bypassed to output common above 200 kHz. This filtering should preferably take place before the shield enters the instrument. If not practical, the filtering should take place as near to the connector as is practical. This solution compromises the greater issue of how to limit the impact of high-frequency fields in instruments.

6.6 COMPLETING THE DIFFERENTIAL AMPLIFIER

The circuit in Figure 6.2 generates a difference signal equal to the input signal difference times some gain. The difference signal can be converted to a single-ended signal referenced to output common by placing the difference signal into a differential stage. This added circuit is identical to the forward referencing amplifier discussed in Section 5.19. The added circuit is shown in Figure 6.5. The circuit details described in earlier sections are not shown for convenience. The feedback resistors used in this stage must be carefully selected and accurately balanced to ensure good common-mode rejection over a wide frequency range. In typical instrumentation designs these resistors are wire-wound to ensure stability and accuracy. The parasitic capacitances across these resistors limit the instrument bandwidth. Special precision metal-deposited resistors are available to handle this application. These resistors have excellent high-frequency characteristics even at 1 megohm that exceed the performance available from wire-wound resistors. Care must be taken in layout to dress these resistors away from other circuits because they can be affected by capacitances above 0.1 pF.

The added differential stage can be used to provide gain. The gain is determined by the ratio of R_1 to R_2 and by the attenuator R_5 and R_6. To reject common-mode signals the attenuator source impedance must be considered. If this source resistance is R_7, then $R_7 + R_2'$ must equal R_2. The overall differential gain becomes

$$G = \frac{R_5 + R_6}{R_6} \cdot \frac{R_2}{R_1} \qquad (6.2)$$

The feedback resistors in this circuit form an input attenuator that affects both normal- and common-mode signals. By design the amplifier has near-zero gain to the common-mode content. If the ratio of R_2 to R_1 is 3:1, this stage can reject 30 V of common-mode signal and still process a normal signal. The gain to normal signals can be set to 1 or any other convenient number. This 30 V of common-mode voltage is any combination of input or input/output common-mode level. However, a penalty is associated with the CMR technique. The differential stage has gain to the internal noise of the IC amplifier. If the feedback attenuator is 3:1, the noise is amplified by the factor of 3. If the input stage has a gain greater than 10, the noise contribution is small. If the input stage has a gain of 1, then differential stage noise dominates.

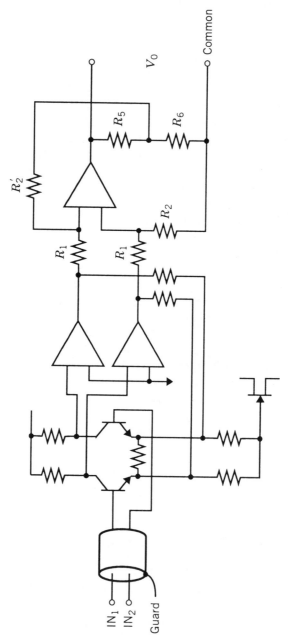

Figure 6.5 The completed differential amplifier.

The rti noise of a balanced differential input circuit is increased because each input transistor contributes noise signal. Since this signal is not correlated, the two signals add by the square root of the sum of the squares. If each transistor contributes 1.3 μV rms, the total noise is 1.84 μV.

6.7 HIGH-VOLTAGE COMMON-MODE REJECTION

The distance between instrument and transducer can be anywhere from a few feet to 3000 ft. Ground potential differences during a test or when caused by nearby lightning activity can often exceed 1000 V. To protect the instruments and to maintain performance under difficult circumstances, it is common practice to require 300 V of common-mode rejection in an instrument. The design approach suggested in the previous section can be changed to provide this level of performance. The power for the input stage can be floating and driven by the input guard. This can be accommodated by using a separate dedicated transformer. The zener supply is still driven by the common-mode signal derived from the two input emitters. This allows the electronics to reject input-type common-mode signals.

The attenuator for the forward referencing amplifier must now attenuate the common-mode signal by 30:1. The common-mode voltage is reduced to 10 V, which still allows the amplifier to handle normal signals. The noise contribution of this stage is thus raised and further voltage stress is placed on the feedback resistors. With more open-loop gain required, the bandwidth, noise contribution, and dc drift place new constraints on the choice of IC amplifier.

The return path for input bias current is now through the feedback attenuator used in the forward referencing section. There is no longer a need to connect the input guard shield through a 1-megohm resistor to output common.

Postmodulation/demodulation techniques use transformers to provide signal isolation and high-voltage common-mode rejection. Care must be taken that carrier signals do not circulate in the system, or there can be crosstalk and even system instability. Another technique involves optical fibers. If the signal is digitized, then fiber optics are insensitive to common-mode processes. Matched photodetectors and special feedback techniques are another way to isolate signals and provide high-voltage common-mode rejection.

High-voltage common-mode rejection techniques place electrical stress on dielectrics, particularly in the connectors. If these stresses last a few minutes, no damage occurs. But if they last for days, then high-voltage dielectric treatment is required. Hence, a different approach to packaging is needed.

A nuclear event causes a significant pulse of electromagnetic energy. Since this pulse couples to all input leads, it represents a common-mode signal with a very steep rise time. The frequency to consider is $1/\tau_r$, which is greater than 30 MHz. This pulse entering an instrument on the input cable can overload many circuits. A 1000-V common-mode rejection test at 60 Hz probably has little relationship to the instrument's ability to respond to this type of pulse. Here is a case where the pulse must be reflected before it enters the instruments if a fast recovery is desired.

Instrumentation is often used to measure the effect of an explosion. An explosion can ionize gases. Ions in motion constitute a current. A pulse of current is a pulse of electromagnetic energy. Care must be taken to avoid common-mode effects which can obscure desired data.

6.8 DRIVEN GUARD SHIELDS

The guard shields used in differential amplifiers are connected to the signal where the signal grounds. In this sense, the input ground drives the shield. The guard potential can be used to drive a power supply common or guard a relay contact. In these examples the guard shield is not driven actively.

In a few cases it may be desirable to reduce input cable capacitance by driving the input shield at approximately the signal potential. This device might be used to extend the bandwidth of a high-impedance source. When the shield moves with the signal, there is no reactive current flow in the input capacitance. This feedback arrangement is shown in Figure 6.6. It is important to recognize that this system has some unique stability problems. At higher frequencies the amplifier has phase shift, which means the input impedance can have a resistive component. If this resistance goes negative, the system might oscillate.

As the cable length increases, it takes time for signals to propagate along the cable and to return. If the signal travels at 150 m/μs, it takes 4 μs to travel 150 m and return. This delay represents phase shift for any sinusoid applied to the cable. At a frequency of 62 kHz the delay is 90°. To reduce input capacitance, the phase shift should be no more than a few degrees.

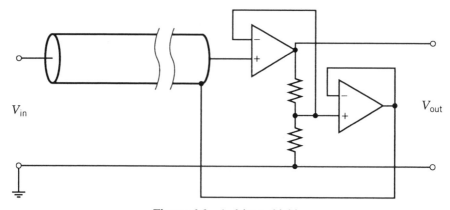

Figure 6.6 A driven shield.

If the input is a step function, the problem becomes more apparent. At first the cable looks like its characteristic impedance, and this attenuates the input signal. A fraction of the step reaches the amplifier, and the shield is driven with this reduced signal. The signal returned on the shield unloads a small fraction of the capacitance. The source knows nothing about this process until a reflection returns to the source. The signal that goes forward upon re-reflection is higher in value. After several round-trips and subsequent delays, the signal may approach its final value. If there is gain in the system, some very interesting oscillations can take place that relate to cable length.

The lesson is simple. If the cable length is known, the gain of the shield driver can be adjusted to reduce reactive loading. It is risky to try this solution on an unknown cable length because an oscillator can easily be produced.

6.9 LOW OUTPUT IMPEDANCES

Feedback circuits used to fix gain also reduce output impedance. Impedance is reduced by the feedback factor. If the open-loop impedance is 10 ohms and the feedback factor is 1000, the resulting output impedance is 10 milliohms. Since loop gain falls off with frequency, the result is that this low impedance rises with frequency. An impedance that rises with frequency is an inductance. This inductance can be seen by noting the response of the amplifier to a square-wave input. The output leading edge will exhibit ringing when a capacitive load is applied to the output. Ringing is the energy decaying in the resonant circuit formed by the source inductance and the loading capacitance.

Capacitive loads can often make a feedback circuit unstable. The capacitor adds phase shift to the open-loop gain, which can cause instability. A cable connected to the output might cause an instability. There are several approaches to handling this problem. The simplest solution is to add a resistor in series with the output so that any capacitive load is not seen directly by the circuit. Often, 10 ohms is sufficient. In applications where loading can affect accuracy, an inductance of about 10 μH can be shunted across the resistor. The inductance keeps the output impedance low at frequencies below 150 kHz. A small inductance can have a series resistance of just a few milliohms.

The output impedance of an amplifier can be very low even though the output current may be limited to, say, 10 mA. Output stages that can deliver high current are often driven by integrated circuits. It can be difficult to stabilize these circuits when feedback is taken directly from the output terminals. The output stage has delay that changes the Nyquist criteria. To overcome this problem, one can take feedback at high frequencies from the IC output. This technique is shown in Figure 6.7.

The output stage bias can be set so that the quiescent current is limited. This reduces unnecessary heating. This bias is set by diodes or resistors in the drive coupling divider. Without feedback this bias places a shelf in the output waveform. This shelf is greatly reduced by the feedback factor. At high fre-

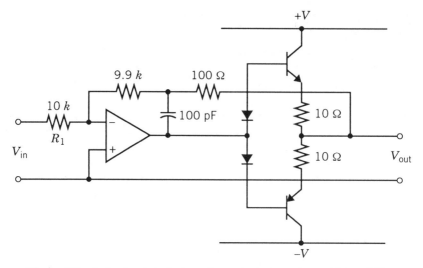

Figure 6.7   A feedback technique for output stages to avoid instability.

quencies where the feedback factor is reduced, the shelf effect introduces distortion. To reduce this distortion, one can change the bias as a function of signal activity. When the output stage is delivering power, the bias level can be actively modified.

6.10   MULTIPLE OUTPUTS

It is common practice to provide multiple outputs in instrumentation amplifiers. One output may deliver higher current to drive long cables, or one may remain unfiltered. In some cases full scale on one output can be set to a lower level to accommodate a tape recorder, for example. If these signals stem from separate amplifiers, a short on one output has no impact on the other. In this sense the outputs are isolated from each other. Often the word *isolation* is not defined, and the user does not get the isolation he or she expects.

Multiple outputs are an invitation for a ground loop. The loop occurs because each output shares the same common. If each output is grounded by the terminating hardware, current will flow in the common connection, because no two ground points are at the same potential and the common signal conductor is not going to short these two points together. Output ground loops often limit system performance. A typical noise specification rto may be 0.5 mV rms. An output ground loop can easily add 100 mV rms noise to a signal.

One solution is to provide a forward referencing amplifier for one of the terminations. See Section 5.20. A second approach is to provide a forward referencing amplifier and its power supply in the main instrument. A separate power supply is expensive if a separate transformer is used. If the power supply common is referenced to the output ground, the ground loop can be

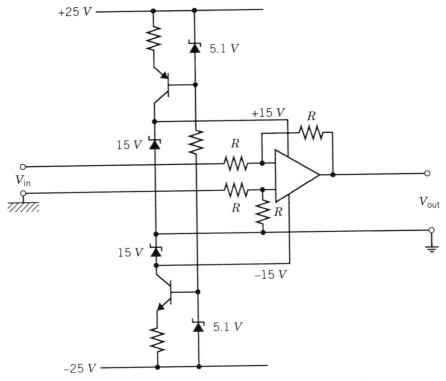

Figure 6.8 A floating output stage that avoids a ground loop.

eliminated. This technique is shown in Figure 6.8. If the supply voltages are 20 V and the regulators are 15 V, the 5 V of "head room" allows the ground potential difference to be several volts. This performance can be given a CMRR specification, and it has no relationship to the two other CMRR specifications discussed before.

6.11 CURRENT SOURCES

A simple feedback circuit can regulate current flow instead of voltage. This technique is used in power supplies and in transducer excitation. The basic circuit compares the voltage drop across a series resistor with a reference voltage. This comparison is the error signal in a feedback system. The circuit is shown in Figure 6.9. If the current has a range of 10 to 100 mA and the sense resistor is 10 ohms, the sense varies from 0.1 to 1 V. The differential amplifier is set to have a gain of 10. The reference voltage can be obtained from a potentiometer placed across a reference zener diode. The circuit is not operational unless an output load is present. The load resistor must be

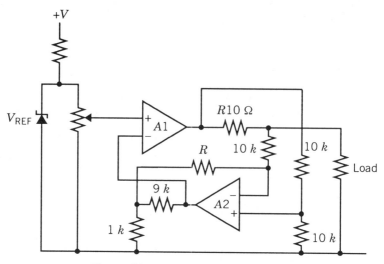

Figure 6.9   A current regulator circuit.

low enough in value so that the output voltage is within the compliance limits of the amplifier.

This feedback system must be checked for stability. One method is to modulate the reference voltage with a low-level square-wave voltage by summing a square-wave signal through a high-value resistor at the potentiometer slider. The signal across the load resistor should be free of significant overshoot and ringing. The amplitude modulation must be small enough so that the output voltage is within its compliance limits. This stability test should be made for low values of load resistance and for both limits of current.

Current regulators used in power supplies frequently regulate current in only one direction. This happens when a series pass element is used. When the current source is modulated by a square wave, its static and dynamic characteristics are tested. The current is changed between two limits, such as 9 and 10 mA. Current reduction is limited. In a test for circuit stability, the output signals must always be within the linear limits of the amplifiers. If the rise or fall times change with signal level, the signals are too large for testing stability. Circuits must be tested for stability at low and high currents for low and high load impedances.

Current sources have high-impedance. Current basically does not change when the load impedance changes. For example, if a load of 1000 ohms is supplied a current of 10 mA, the output voltage is 10 V. If the load is changed to 500 ohms, the voltage may drop to 5.005 V. This means the current changed from 10 to 10.01 mA. The output impedance is simply $\Delta V/\Delta I = 0.5$ megohm. Output impedances of 10 megohms are quite practical.

This method of testing requires very accurate resistors and voltmeters and is not easily made. Further, it only tests for high impedances at dc. A dynamic testing method is discussed in Section 7.20. This test measures ac and dc performance without the need for accurate testing equipment and displays system stability.

6.12 CURRENT LOOPS

A single loop of wire can be used to supply power and monitor parameters at a remote point. The loop can supply voltage or current to a remote point and sense current or voltage returned from that point. If the source of power is floating (derived from a separate transformer or transformer coil), the loop can be grounded at some external point.

The simplest loop circuit involves a resistive transducer that measures temperature, pressure, or vibration. If the source of power is a constant current, the voltage at the transducer measures transducer resistance. As the resistance varies, the voltage at the current source measures the parameter of interest. This simple circuit cannot compensate for the resistance in the conductors if it varies with temperature. This voltage drop is an error that must be considered.

If a constant-current source is involved in sensing vibration, the transducer signal at dc is not important. In this application the signal can be sensed at the source of current and decoupled by using a capacitor. The voltage drop in the leads is of no concern. The signal returned from the transducer can be low impedance if the circuit is active or simply the impedance of the transducer element if the circuit is passive. The *RC* time constant must be set so that the low-frequency response is acceptable.

Consider a constant-current source operating an active circuit. The signal returned on the loop is proportional to the apparent impedance of the circuit plus the power supply voltages (zeners) and any line drops. If the signal of interest is set to a known value, the voltage returned can be used as a reference level. Variations around this reference level measure the signal of interest. This reference value can be reset periodically. The reference value compensates for variations in power supply and line voltage drop. The active circuit can respond to a variety of signals for this resetting process. As an example, the current source can be modulated by a carrier or simply turned off momentarily. The circuit sensing this command shuts off the signal for a short interval. Capacitors across the power supply maintain voltage during this operation.

A single loop can also be operated from a fixed-voltage power supply. A series sense resistor measures the current flowing in the loop. The active circuit at the transducer can convert the signal of interest to a loop current. This current can then be measured across the sense resistor at the voltage source. Any fixed current flowing in this resistor produces an offset that can be compensated for in the subsequent electronics.

In general, current loops are not recommended where high-frequency operation is required. The reactance of the line parallels the current source and attenuates the signal. A 1000-ft loop with 20 pF/ft capacitance looks like 0.02 μF, which represents a reactance of 10,000 ohms at 800 Hz.

Current loops are often from 0 to 20 mA. If the impedance involved is 1000 ohms, the compliance voltage is 0 to 20 V. This range of signal needs to be buffered but not amplified.

6.13 ACTIVE FILTERS

Low-pass filters are often incorporated into instrumentation amplifiers. These filters can be linear phase, Butterworth, Chebyshev, elliptic, or special characteristic, depending on application. Filters can be as simple as third order or as complex as eighth order. Cutoff frequencies can be selected from as low as 1 Hz to perhaps 30 kHz. The selection can be in 1/3-decade steps, in one, two, five steps, or at specific frequencies. With so many variables the complexity and cost can vary considerably.

Cost often dictates that active filters are designed into the output circuitry. This means that all of the gain provided in the instrument precedes the filtering. The main reason for this approach involves the drift and offset of the filtering electronics. Many active circuits are involved, each contributing offset errors. If gain follows filtering, these errors must be proportionately lower, which can be difficult to achieve. Filters with cutoff frequencies below 10 Hz must operate at relatively high impedances or the capacitor sizes become unreasonable. At high impedances bias currents can introduce further dc errors. Since bias current varies with temperature, these errors are not easily handled.

Filters are intended to remove spurious signal content without disturbing the desired information. Filtered data are much easier to read and interpret. If there is a limited bandwidth requirement, filters are often used to attenuate (obscure) unwanted power-related coupling. The proper way to handle power-related interference is to remove the interference before using a filter. If the interference increases to the point of overload, the user may be unaware that the data have been compromised.

In an ideal world, filtering should be followed by gain. When an unknown signal is involved, the form factor can be quite high. (Form factor is the ratio of peak to average signal level.) Consider the case where the noise doubles the peak signal level. An 8-V signal has 16-V peaks, above the limits of the instrument. The instrument clips these excessive signals and recovers within microseconds. Unfortunately, filtered clipped data are not valid. If the clipping is unsymmetrical, one of the errors is a level shift. If the filtering was followed by a gain of 5, the 16-V signal would only be 3.2 V. The filtered data at the output would be an undisturbed 8 V.

6.14 FILTER RESPONSE

All filtering processes introduce time delay. A Bessel filter offers the same delay for frequencies in the passband and for frequencies well beyond the cutoff frequency. If the cutoff frequency is doubled, the delay is halved. The

Bessel response is used in oscilloscope design because it has essentially no overshoot to a step function input. It is a good filter type to use in instrumentation because it introduces no overshoot or ringing, which can affect the data. For data to be analyzed by computer this sort of transient character may be of no concern as long as clipping or overload does not occur.

Sinusoidal delay is a function of filter type as well as cutoff frequency. Except for the Bessel filter the general low-pass filter does not provide constant delay for sinusoids in the upper portion of the passband: the sharper the knee, the greater the departure from linear phase. It often makes mores sense to consider group delay. If the signal being processed is a square wave, the group delay is approximately the time to the 50% point on the leading edge.

Bessel filters have a very soft knee around the cutoff frequency. Adding orders to the filter does not sharpen the knee. Butterworth filters provide the sharpest knee possible for a flat response in the passband. The penalty for this sharpness is overshoot to a step function input. If slight amplitude variations are permitted in the passband, the knee can be further sharpened but at the expense of even more overshoot. Amplitude ripple in the passband is provided by Chebyshev filters. Filters with ripple in the stopband or in both pass- and stopbands are called elliptic filters, and further increase the overshoot and ringing problem. Overshoots in excess of 40% are not uncommon.

Computers can be used to synthesize filters that control phase and amplitude in special ways. For example, a filter can be designed with Bessel character to 0.9 times the cutoff frequency, essentially no overshoot, and a reasonably sharp knee. This type of filter has no standard name, but it can be described mathematically in terms of pole and zero location. Testing can be handled by noting overshoot, 3-dB points, and attenuation at several stopband frequencies.

6.15   THE ALIASING ERROR ISSUE

When data are sampled, strange answers might be obtained. Looking outside once every 25 hours, one sees the sun advancing 1 hour every day. In 24 days the sun would make one complete revolution. This is obviously not the length of the day. A similar error occurs if the sampling occurs once every 23 hours. This same type of error can occur in sampled data. The only way to ensure that this cannot happen is to sample more often. The rule is simple: more than two samples must be taken per cycle at the highest frequency of interest. For this reason filters with a flat frequency response to some upper frequency and no response above that frequency are desirable. At frequencies greater than twice the sample rate, there should be no signal content, or errors will be introduced. Raising the sample rate to limit aliasing errors has an impact on system cost and performance because there are more data to store and analyze.

6.16 ACTIVE LOW-PASS FILTER DESIGN

Filter characteristics are described in the complex frequency domain by the location of poles and zeros. Pole pairs can lie in the left half-plane, or single poles on the negative real axis. The pole pairs are always symmetrical about the horizontal axis. The response of a filter with a given pole/zero configuration can be determined graphically. Points along the vertical axis represent frequency with dc at the origin. Draw lines from the frequency selected to each pole and zero. Multiply the lengths of these lines and form products $P_1 P_2 P_3 P_4$ and $Z_1 Z_2$. Draw lines from the origin to each pole and zero. Form the products $p_1 p_2 p_3 p_4$ and $z_1 z_2$. The response at any frequency is $p_1 p_2 p_3 p_4 Z_1 Z_2 / z_1 z_2 P_1 P_2 P_3 P_4$. See Figure 6.10.

Draw a line from the origin to either pole of a given pole pair. The length of this line is the natural frequency of the pole pair. The damping factor is

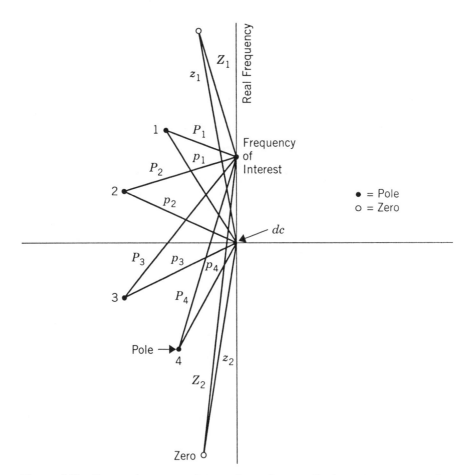

Figure 6.10 Geometric construction to determine amplitude response versus frequency of a filter, given pole and zero locations.

the cosine of the angle between this line and the real axis. When the angle is zero, the pole pair coalesces and the resulting two-pole section is said to be critically damped. For a second-order filter (single pole pair) the damping factor provides no overshoot to a step function input. For damping factors less than unity there is always an overshoot.

The location of the poles for various filter types is given in the literature. For Butterworth filters the -3-dB point is given for $\omega = 1$. In other filter types the books give pole locations that may place the knee of the filter at $\omega = 1$.

Butterworth filters are characterized by pole pairs lying on a circle in the left half-plane. A third-order filter has one pole on the horizontal axis and two poles on the 45° radii. In a Bessel filter the poles lie in the left half-plane approximately on a circle whose center is in the right half-plane. A Chebyshev filter has poles located on an ellipse. An elliptic filter has a group of left half-plane zeros located near the vertical axis at frequencies in the stopband.

A common approach in active filter design is to allocate one operational amplifier to each pole pair. In an odd-order filter one of the operational amplifiers can be used to provide a third-order section. Cascading two operational amplifiers allows a fourth- or fifth-order filter to be used.

A second-order filter circuit is shown in Figure 6.11. The natural frequency is

$$\omega_n = \frac{1}{\sqrt{R_1 R_2 C_1 C_2}} \qquad (6.3)$$

The damping factor is

$$\sigma = \frac{ab}{2(a^2 + 1)} + \frac{ab}{2}(1 - G) \qquad (6.4)$$

where $R_1/R_2 = a^2$, $C_1/C_2 = b^2$, and G is amplifier gain.

In filter design it is far easier to use standard capacitor values and adjust the damping and natural frequency by selecting resistor values. Standard

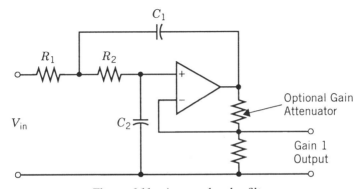

Figure 6.11 A second-order filter.

accurate capacitor values are often limited to the decade values. Paralleling capacitors makes practical ratios of 2:1 and 3:1. The lowest damping factor occurs for equal resistor values. If $C_2/C_1 = 2$, the lowest damping factor is 0.707. A damping factor of 0.8 can be set by changing the resistor ratio.

For damping factors lower than 0.707, gain can be provided to drive the feedback capacitor. A simple way to obtain a small gain without adding another integrated circuit is to raise the gain of the filter circuit itself. The filter capacitor is connected to the amplifier output. If the filtered signal is taken from the feedback divider, the stage gain is still unity. Care must be taken not to load this point because any loading changes the damping factor. For a gain of 1.2 the lowest damping factor possible is 0.566. For damping factors below this value the capacitor ratio should be increased. Gains greater than 1.2 limit the available full-scale output voltage.

In an odd-order filter the odd pole can be handled by adding an RC circuit ahead of the second-order filter section. The value of R in the RC circuit should be about one-tenth the value of R used in the main filter. The selection of component values is an iterative process, because the equations that must be solved are not linear. It is easiest to associate the odd pole with the highest damped pole pair. In a cascaded arrangement the least-damped pole pairs should be last, to stop ringing from overloading the filter.

Active filters can be designed with state variable techniques. This approach is used in elliptic filter design where complex zeros of transmission are required. The state-variable approach uses more active elements than the method just described. It has the advantage of reducing the sensitivity to variation in individual element values. The use of digital-to-analog converters (DACs) makes it practical to digitally program this type of filter. This approach is powerful because the poles and zeros can be moved to provide different cutoff frequencies and to modify filter type. Now the designer and the user are no longer restricted to specific cutoff frequencies or standard filter types.

6.17 CHARGE AMPLIFIERS

A charge amplifier is formed by using capacitors for operational feedback elements as shown in Figure 6.12. The voltage gain of this circuit is the ratio of impedances or simply C_2/C_1. The ratio of voltage to charge on any capacitor is simply the capacitance, or $C = V/Q$. In this feedback circuit any charge generated at the input must appear on the feedback capacitor, which determines the output voltage. If the input charge is $V_{IN}C_1$, the output voltage is Q/C_2 or $V_{IN}C_1/C_2$.

A piezoelectric transducer is simply a crystal that responds to a force between two faces by developing a charge. From Newton's law, $f = ma$, the voltage at the transducer is a direct measure of acceleration. A convenient way to convert acceleration to a low-impedance voltage is to use the crystal as capacitor C_1 in Figure 6.12. Any charge generated by the crystal is converted

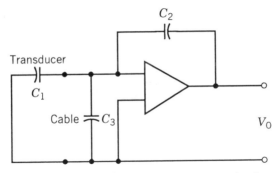

Figure 6.12   A simple charge converter circuit.

to a voltage at the amplifier output. The charge is removed by the circuit with the result that the voltage is zero.

The crystal transducer is usually located at the end of a cable. The cable capacitance appears between the summing point and circuit common. Since there is no voltage at this summing point, the cable capacitance does not affect the charge converter gain of the circuit. Cable capacitance does affect the noise gain of the stage because the noise gain is the ratio of $(C_3 + C_1)/C_2$. It also reduces the amount of feedback available to control gain, bandwidth, and linearity.

A capacitor feedback arrangement converts the charge appearing on the transducer to voltage, and thus the circuit is called a charge converter. The circuit is limited at low frequencies by the leakage resistances of the cable and transducer. Typically these leakages allow transducers and cables to function down to about 0.5 Hz. If the RC time constant is 2 s and C is 500 pF, then R is 4000 megohms. The circuit itself has limitations because some input current must still flow. When FETs are used at the input, bias current may be a few nanoamperes. If this current has no path to follow, the circuit will saturate. The path is usually a high-valued shunt resistor placed across the feedback capacitor. In many designs an internal circuit must be provided to supply this current. This adjustment also serves to zero the output of the charge converter circuit. If not done, overload cannot occur.

A typical charge amplifier has a series of gain settings. Gain can be provided by reducing the feedback capacitor, or voltage gain can follow the charge converter stage. The gain following the charge converter can be set to accommodate transducer calibration. If the transducer calibration is in picocoulombs (pC) per g, then this setting can allow the output voltage to indicate acceleration in engineering units. For example, each volt could represent 2 g's of acceleration.

The gain of a charge amplifier has units of mV/pC. If a transducer has an output of 5 pC/g, then 10 g's generates a charge of 50 pC. If 10 g's is to provide a full scale of 10 V, the gain must be 10 V/50 pC or 0.5 V/pC (500 mV/pC). If the input feedback capacitor is 1000 pF, the charge converter generates 50 pC/1000 pF or 50 mV. The voltage gain following must then be 200.

The noise referred to the input for a charge converter might be 0.005 pC rms in 100-kHz bandwidth. At a gain of 500 mV/pC the noise is 2.5 mV rms. The peak noise might be three times this value, or 7.5 mV. Compared to a 10-V peak output, this signal-to-noise ratio is satisfactory. The $1/f$ noise is likely to be high in a charge amplifier as the feedback reactances are quite high in the lower-frequency range. The rms values measured are a function of the integration time of the recording voltmeter. For noise to be measured down to 0.5 Hz, the voltmeter should be able to measure signals as low as 0.1 Hz. Digital voltmeters that analyze sampled data can have this required bandwidth.

The smallest practical feedback capacitor is about 200 pF, which represents a reactance of 1592 megohms at 0.5 Hz. Board and component leakage resistances should be kept at least 10 times greater than this value. This low value of capacitance can provide high charge converter gain, which optimizes the signal-to-noise ratio of the instrument.

The low-frequency response of the charge converter can be controlled by placing a resistor across the feedback capacitor. This resistor also serves as a path for input bias current. For the 3-dB point to remain constant, the resistor should be changed for each feedback capacitor used. If the largest feedback capacitor is 0.01 μF, a 31.8-megohm resistor provides a 1-Hz, 3-dB point. When the capacitor is 1000 pF, the resistor should be 318 megohms. An output voltage divider can be used to set the time constant. This allows one moderately high value resistor to be used with all capacitors. The circuit is shown in Figure 6.13. At 200 pF the resistor would have to be 1590 megohms, not an easy value to obtain in small size.

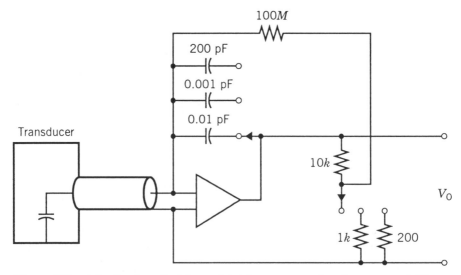

Figure 6.13   A feedback method for maintaining a constant low-frequency −3-dB response.

6.18 PYROELECTRIC AND TURBOELECTRIC EFFECTS

Charge is developed in insulators when there is a temperature gradient (pyroelectric) or when the materials are mechanically stressed (turboelectric). This phenomenon can occur in cables as well as in the transducer itself. Pyroelectric effects are apt to appear as a steady current flow, which, in a charge converter, can block the converter circuit unless an adequate dc path is provided. The shunting resistor across the feedback capacitor can provide such a path. When this current is flowing, the converter has a voltage offset. If the shunting feedback resistor value is too high, the converter can overload. This phenomenon often forces limits on the low-frequency response of charge instrumentation.

Special low-noise cable must be used between the transducer and the instrument. The cable shield is conductively treated so that charge is not developed when the cable is flexed or stressed. This treatment eliminates turboelectric phenomena in the cable.

6.19 LEAKAGE CAPACITANCES IN CHARGE CONVERTERS

Consider a charge converter with a feedback capacitor of 200 pF. If the accuracy is to be held to 1%, the capacitance must be held to 2 pF. Hence, care must be taken in mounting the capacitor so that any parasitics present are constant. The gain will be a function of component value and parasitic capacitances.

Circuits with any signal gain near the converter can introduce significant gain error by reactively coupling to the input. If there is a gain of 1000 following the charge converter, the reactive coupling to this signal must be less than 0.002 pF—a small value indeed. Thus, the charge converter must be built inside an electrostatic enclosure, or large gain errors can result. This implies that capacitor switching circuits must also be inside the enclosure.

Experience shows that millivolts of power supply ripple can couple signals into the input of a charge converter. The entrance is not through the integrated circuit but capacitively to the feedback capacitor or input summing point. Input leads must be kept short and closely dressed. Power supply leads must be well filtered before being brought into the enclosure.

6.20 IMPULSE AND IMPACT TESTING WITH CHARGE AMPLIFIERS

Impulse testing involves signals with steep leading edges, which are likely to be large and require low gain settings for the charge converter. A typical low-gain feedback capacitor is 0.01 μF. A step input charge requires a step output voltage. The current supplied to the feedback capacitor depends on the rate of change of voltage. If the charge converter signal rises to 10 V in 20 μs, the current required is 50 mA, beyond the capability of most standard integrated circuits. A driver stage must be supplied if this performance is required. The driver must be able to supply current in both directions. Adding a driver stage makes it necessary to stabilize the circuit at all gains. Stability must be checked for small and large signals.

Most piezoelectric transducers exhibit resonance at some upper frequency. If the resonance has a high Q, the transducer can easily generate signals that can overload an amplifier unless filtering is provided in the instrument. Impulse testing is quite likely to excite such resonance. If the instrument filter is in the output stage, the data will probably be clipped before filtering, which can destroy the test results. Filters should be placed ahead of the last stage of gain to avoid this class of overload.

6.21 RECOVERY PHENOMENA IN CHARGE AMPLIFIERS

An ac system implies coupling capacitors or transformers. Coupling components that allow frequency response at frequencies below a few hertz must involve long time constants. When energy is stored in these coupling components, it takes time for it to dissipate. During dissipation an output signal must occur. This energy might be stored during overload for certain type of signals or during turn on. If several components are involved, the recovery process can be nonlinear and take several seconds. If the recovery involves multiple overloads, data may be lost. For this reason it is desirable to minimize the number of coupling capacitors.

A charge converter involves at least three capacitors: The transducer itself, the feedback capacitor, and at least one coupling capacitor out of the charge converter. The stages that follow can be dc-coupled. If they are ac-coupled, the frequency response can be correct but the response after a step input could ring for many seconds. There is no way to avoid some recovery response, but it can be designed to be gentle, without excessive ringing.

The transducer, the feedback capacitor, and shunting resistor in the charge converter circuit constitute a single pole of transmission with a zero of transmission at dc. If the resistor is tapped and a capacitor shunted to common, the response can be corrected (made flatter) at low frequencies. This addition makes the response second order. The coupling capacitor out of the converter can then add a third pole. By proper design the high-pass filter that results can be a third-order Butterworth equivalent with good recovery characteristics. This approach allows a flat frequency response, a sharp cutoff characteristic, and a good compromise on recovery and settling time. To achieve the same flat low-frequency response without complex poles of transmission requires larger components and, hence, a longer recovery time. This circuit is shown in Figure 6.14.

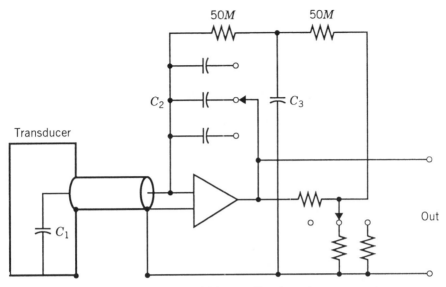

Figure 6.14   A third-order high-pass filter in a charge converter.

6.22   DIFFERENTIAL CHARGE AMPLIFIERS

A crystal transducer is basically a single-ended device. For a crystal transducer to be balanced, two crystals must be placed in series and the midpoint bonded to the shield. The amplifier, then, must be two accurately balanced charge converters. Balancing charge converters is not simple, and increases the cost significantly.

A balanced transducer is desirable because of problems associated with bonding transducers to a structure. Balancing the transducer is not necessarily the way to treat this type of instrumentation problem. When a transducer is mounted in a housing, the housing is necessarily part of the input shield. To be effective, the input shield must connect to one side of the signal. If this housing were to contact the structure under test, the input circuit would in effect be grounded. Grounding the output of the charge amplifier causes a ground loop and adds significantly to the noise, since noise current would flow in the input shield. One solution is to insulate the transducer case where it is mounted. Insulation adds another degree of freedom to the mechanical structure and can be undesirable.

Another solution is to make the charge amplifier differential, although the input need not be balanced, but the input and output can be grounded and the ground potential difference removed as a common-mode signal. The general nature of the solution is shown in Figure 5.11. The first section becomes the charge converter, and the second section adds gain. This second amplifier can be similar to the designs in Figures 6.3 or 6.5. Charge instruments built this way allow the transducer to be floating or grounded without reconfiguring the instrument.

6.23 CALIBRATION OF CHARGE AMPLIFIERS

Calibration techniques should include cable, transducer, and instrument where possible. Piezoelectric transducers and their instruments provide some interesting problems and compromises. In strain-gauge instrumentation the gauge itself can enter the calibration. In a charge instrument, calibrating through the transducer is not as easily accomplished.

A capacitor connected to the summing point can be used for charge insertion. Since a charge amplifier is an ac device, the signal might be a sinusoid from an external generator. Unfortunately, any external source has its own ground, and it should not be connected to every instrument at the same time. One way to interface an external generator is to differentially isolate the generator in each instrument. This has the added advantage of controlling the generator loading. Without buffering, the generator would see a large group of parallel capacitors. The generator signal cannot be brought into the input shield region because very small capacitances can couple signal to the input. Instead, the generator should be decoupled from the differential isolator external to the charge converter shield. This disconnect is best handled by a relay.

Calibration should serve to show that the amplifier is present, that the gain is correct, or that the transducer, cable, and amplifier are correctly set and functional. Calibration using one signal level implies that the gains must be changed to accommodate this signal source. This means that the gain setting in use is not actually checked. The alternative is to calibrate each instrument in sequence and program the calibration signal for each channel of instrumentation.

Charge insertion verifies that the charge instrument is operational. Unfortunately, the cable or the transducer can be disconnected and the calibration results do not change. For all this trouble, this form of calibration does not verify that the cable or transducer are connected.

A calibrated hammer can be used to shock-test an entire system. The response can be checked before and after a test to determine if the system is still repeatable and intact. If the mechanical system changes as a result of the test, this approach may not be viable. The test can always point out whether transducers and cable are still connected.

A transformer can be used to insert a signal into the input cable. This voltage is in series with the cable and transducer capacitances and defines a charge input. This calibration procedure does not verify the gain of the charge amplifier as seen by the transducer, but it does verify that the entire signal path is intact. The transformer primary must be shorted when the calibration signal is disconnected. The signal on the primary of the transformer must be referenced to the signal common of the charge converter circuit.

6.24 CAPACITORS AND CHARGE AMPLIFIERS

The feedback capacitors determine gain accuracy. The capacitors should be stable and have low dielectric absorption. In most instruments the selection of capacitor type is a compromise. The temperature coefficient should be near zero around room temperature. The best capacitor dielectric is polystyrene, although polypropylene is quite satisfactory.

In instruments that use DACs to control gain, the value of the feedback capacitors can be absorbed in the gain setting. This allows the designer to use 5% capacitors and still meet a 0.5% accuracy specification.

The recommendation made earlier suggests dc amplification after the charge converter. When this is not done, large coupling capacitors may be required to maintain a required low-frequency response. Large capacitors in the signal path are apt to exhibit nonlinear characteristics at low frequencies. This phenomenon may be very difficult to detect unless a digital signal analyzer is used. In applications where low-frequency data are important, this aspect of performance should be carefully checked.

CHAPTER 7

TESTING OF AC AND DC AMPLIFIERS— STRAIN-GAUGE ISSUES

7.1 COMPARISON METHODS

The basis of many measurements is comparison. Small departures from the expected result can be observed by subtraction. The basic technique is to form a null between an input driving signal and a signal under test. Figure 7.1 illustrates this technique for active devices. If the device has a gain G, then a precision attenuator with loss $1/G$ is placed ahead of the input. If the device under test (DUT) is accurate, the generator signal and the output signal cancel. This generally assumes an amplifier with a dc response, configured to have a negative gain. Any difference or error is measured as a small signal at the midpoint of R_1 and R_2. Because $R_1 = R_2$, only half the error is sensed. These two resistors should be precision metal film or wire-wound precisely matched in the circuit. In this test circuit the generator source impedance produces no error. The output impedance of the DUT adds to the impedance of the summing resistor. If R_1 and R_2 are 10,000 ohms, the output impedance should be under 1 ohm to keep the summing error below 0.01%.

The attenuator can consist of an external 5-decade resistor box R_0 and a precision four-terminal wire-wound 10-ohm resistor R_3. This allows the attenuation to be set to the expected gain of the DUT. If the gain is 100, the resistor box should be set to 990 ohms. A four-terminal resistor has two points on the connecting leads where the resistance is exactly 10 ohms. These sensing points must be lightly loaded by the input impedance of the DUT. The connections and contacts to the other terminals of the precision resistor should be robust. If gold-plated connection terminals are available, they should be used.

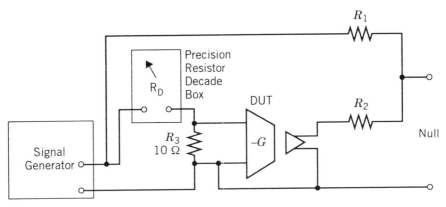

Figure 7.1 A simple null test circuit.

Contact resistances should be below 1 milliohm. The leads to the resistor box should be at least two parallel No. 14 conductors.

Precision matching of resistors R_1 and R_2 can be accomplished by measuring the midpoint voltage. When the resistors are swapped, the midpoint voltage should not change. A potentiometer in series with the smaller resistor can be adjusted to get an exact match.

Current in the summing resistors must flow from the amplifier output through the signal generator. This current path includes wiring that connects the input attenuator. All of this wiring must be low impedance to keep errors below 0.01%. A 1-ohm error is 0.01% for the amplifier output impedance. If the gain is 1000, the wiring at the input must provide connections that are kept below 0.001 ohms. This takes careful construction. The device under test must be single ended for this summing process to function. Any input/output common lead switching must also consider the path for output current flow.

The signal generator can be sine wave, square wave, or triangular wave. Each signal type has its application. If the DUT were perfect, the null would be exact for any type of signal at any frequency. The summing resistors should be dressed away from the output and signal source to limit the parasitic capacitances.

Several refinements are suggested. The sense point between the two summing resistors can be moved 0.1% by adding a small circuit in series with one summing resistor. This circuit consists of a potentiometer (1000 ohms) across a small resistor (10 ohms). A switch can be used to select which summing resistor is used. The sensing point is the slider of the potentiometer. This technique provides additional resolution not available by adjusting the external decade resistor box. It also allows direct reading of small gain errors. Diodes can be placed back-to-back from the summing point to common to clip large signals, necessary for square-wave testing. During transitions large signals can overload the monitoring instrument. It is desirable to switch load resistors onto the output of the DUT, to allow gain and linearity checks when the instrument is delivering full current.

Null testing does not rely on the accuracy of the signal source. A sine wave signal can have 2% distortion and the results are not affected. The amplitude of the testing signal need not be precisely set to obtain results.

7.2   NULL SIGNALS

Ideally the null point has no signal. Consider the test of a dc instrument using a 10-Hz sine wave. If the DUT is wideband, the phase shift should be near zero at this frequency. The null should be observed as a Lissajous pattern with the horizontal axis of an oscilloscope connected to the signal generator. If the output signal of the DUT saturates or clips, the null pattern appears as in Figure 7.2. The clipping points are a measure of the peak output signal. Departures from a best straight line provide a measure of linearity, which is independent of offset, drift, noise, generator quality, or gain value. If a load is placed on the output of the instrument, changes to linearity or peak output are immediately apparent. If the peak-to-peak error is 3 mV, the actual error is 6 mV peak-to-peak, a linearity figure of +0.03% for a 20-V peak-to-peak output signal. The attenuator setting to obtain a best null is, by definition, the gain. This measure is independent of offset and linearity errors.

7.3   OUTPUT IMPEDANCE

A null pattern should first be obtained for the no-load condition. When the load is added, the null pattern changes slightly, indicating a new gain. If the gain changes 0.1%, the output impedance is 0.1% of the load resistance value. If

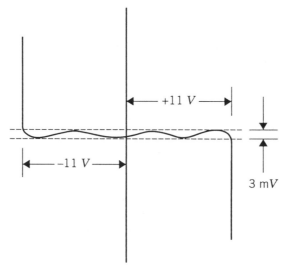

Figure 7.2   A typical null pattern for a sinusoidal input.

the load resistor is 100 ohms, the output impedance is 0.1 ohm. This technique is valid only at frequencies where DUT phase shift is near zero.

This test includes any wiring from the output terminals to the testing device. For measurement of the output impedance of an instrument, excluding wiring, the load resistor must be located at the instrument connector. For evaluation purposes if the resistor is placed at the internal sense point, all external resistance can be removed.

7.4 SQUARE-WAVE TESTING

The phase shift associated with sine wave testing precludes obtaining a measure of gain or linearity at high frequencies. In a 100-kHz instrument the phase shift might be several degrees at 100 Hz, and the null signal introduced by phase shift can far exceed the linearity error. Null measurements using square waves eliminates the phase shift issue. Gain is observed by obtaining the best straight line, as before. This test also quickly provides settling time information. If the output signal goes from -10 to $+10$ V, the instrument must recover to within 20 mV to settle within 0.1% of final value. At the null point this means settling to within a 10-mV band.

The gain should be checked for various square-wave repetition rates. A difference in gain indicates a problem. If the gain for 3-kHz square waves is lower than the gain for 100-Hz square waves, this might indicate a lack of high-frequency loop gain in some portion of the amplifier.

Linearity cannot be checked directly, but it may be inferred. If the gain remains unchanged after settling for all square-wave input signal levels, the linearity is within the gain error out to a frequency of $\frac{1}{4}f_c$. Any test should include square-wave repetition rates from 10 Hz to perhaps 1% of the upper -3-dB point to verify that the gain does not change with frequency. This test of dynamic linearity is powerful, although a numerical measurement value is not readily available.

7.5 TESTING OF AC AMPLIFIERS

The phase shift through an ac amplifier is zero at some mid-band frequency. Null testing with sine waves can only be made at this frequency. When square waves are used, the sag phenomenon associated with an ac amplifier makes it impossible to achieve a total null. A better null pattern can often be obtained using a triangular wave. This setting is the gain around mid-band. A typical null pattern is shown in Figure 7.3.

A charge amplifier can be tested for gain when a known capacitor is used at the input. The input charge is simply the input voltage times the input capacitance. The capacitor must be well shielded and connectorized to avoid stray electrostatic pickup.

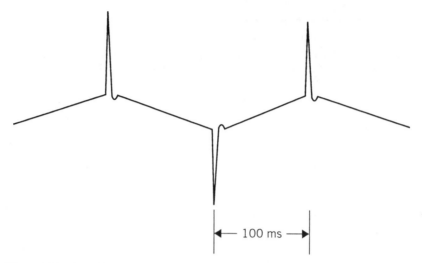

Figure 7.3 A null pattern using square waves to measure gain on an ac system.

7.6 HARMONIC DISTORTION

Harmonic content is generated by nonlinearities in an amplifier. A sinusoidal signal should be used as an input signal. Any distortion appears at multiples of the fundamental driving frequency. This content can be measured by filtering out the fundamental of the driving signal and measuring any remaining signal content. The signal generator must be reasonably free of harmonic content or this method fails. This measure cannot be related to linearity because the phase relation between harmonics is unknown. The signal generator should be checked for distortion before accepting data from the analyzer. The analyzer can measure one rms number (total harmonic distortion) or provide information on the content at each harmonic.

7.7 INTERMODULATION DISTORTION

Unwanted signals can be generated in an amplifier when two signals are processed simultaneously. Any signal content at the sum and difference of the two signals is intermodulation distortion. This measure must be made across the band of the amplifier. Any distortion figure should reflect the worst case observed.

In dc amplifier testing, one of the two input signals can be dc. When an ac signal is superposed, the harmonic content can vary as a function of offset. This content can be measured with a distortion analyzer. The sum of the two signals must be within the compliance limits of the instrument, or the measure is invalid.

7.8 SLEW RATE AND BANDWIDTH

An amplifier may have a bandwidth of 100 kHz and an output capability of 20 V peak-to-peak. When the amplifier processes a sine wave at this upper frequency, the voltage changes most rapidly at the zero crossing. This maximum rate of change is $2\pi f V_p$, or in this case 6.28×10^6 V/s. This rate of change is called the slew rate. There is no guarantee that an amplifier will process a full-scale signal at maximum bandwidth, although this is desirable. Unless specifically stated, these two specifications are unrelated.

Maximum output at maximum frequency may not be achieved for all gains. At unity gain the input stage must operate over its maximum dynamic range, which can be a difficult design problem.

Bandwidth should always be measured using small signals. If the bandwidth varies with signal level, slew-rate limiting is undoubtedly present. In the preceding example, if the slew rate were only 0.628×10^6 V/s, the bandwidth would have to be measured at a 1-V peak signal level.

Square-wave testing provides a good indication of slew rate. If the rise or fall time varies with signal level, slew-rate limiting exists. The slew rate is approximately the rise or fall time for a full-scale signal. In a linear system the rise and fall times for a square-wave input should be constant.

Slew rates can differ for positive- or negative-going signals, usually because current can flow in only one direction in internal transistors. If a capacitor stores a charge, an active element may have no way to remove it. The circuit is in overload until the charge drains through some passive circuitry. Transient recovery from slew-rate overload can differ considerably from normal transient response. During slew-rate limiting any indication of ringing on the leading or falling edge should be viewed with suspicion because this is a sign of trouble.

Slew-rate limiting is a form of rectification. In a dc amplifier, it usually causes a dc shift. If the signal cannot reach its positive full scale, the signal has a negative offset, which can be detected with a simple dc analog voltmeter. In some cases slew-rate limiting is symmetrical and little or no offset is present. The offset may only be a few millivolts for a 10-V signal, which might easily go undetected.

During slew-rate limiting an amplifier is actually in overload, which means that loop gain is reduced. When an amplifier is processing a midband signal and rf interference is producing slew-rate limiting, there may be lost gain and, thus, unwanted distortion may be present. Hence, rf content, particularly on input stages, must be limited. A crude test to see if rf filtering is adequate is to operate the equipment out in the open. When the hand is placed near the input stage, the output should not offset. A good design should have no hum pickup either.

During slew-rate limiting the output can often show signs of instability, which might appear as a slight ringing along the leading edge of a square wave. If an internal stage is the culprit, then reactive load tests may actually obscure this phenomena.

7.9 SAG AND STEP RESPONSE IN AC SYSTEMS

The simplest decoupling circuit is a series RC circuit. The circuit, shown in Figure 7.4, is characterized by its -3-dB point. If $R = 1$ megohm and C is 1.0 μF, the -3-dB point is at 0.159 Hz. How long does it take the signal to sag 1% and 10% for a step function input? The answers are 10 ms and 105 ms. The product of R and C is called a time constant. In the example $RC = 1$ s and the sag is about 10% in 0.1 s. A 1-Hz sine wave is attenuated by less than 0.02%. It should be clear that it is very difficult to limit sag by adding low-frequency response. As indicated, it takes 10 time constants to limit the sag to only 10%.

A transformer has no response at dc. When a square wave is processed through a transformer, some sag usually occurs. The time constant relates to source impedance, coil resistance, and magnetizing inductance or R/L. If the transformer were ideal, a step function input would have no sag. Faraday's law states that the rate of change of flux is equal to the voltage applied. If the core does not saturate, the voltage on any coil has the same waveform as the driving coil.

When a signal consists of a step function and some superposed high-frequency content, the peak-to-peak signal can be significantly different for a dc amplifier than for an ac amplifier. A step function will sag in the ac system, allowing a superposed signal to be processed. Without sag the head room for a superposed signal is reduced unless the added signal is unidirectional.

Every coupling element that can store energy at signal frequencies must return this energy to the system when the signal is removed. (Energy stored by the power supply or operating potentials is not returned to the circuit until power is turned off.) This energy return results in output signals that can last many seconds. The only way to avoid this phenomenon is to use dc instrumentation.

When a sinusoidal signal is impressed on an ac-coupled system, the output consists of that sinusoid plus some transient content. This transient must exist

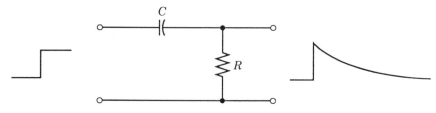

Figure 7.4 A simple decoupling circuit.

if the steady-state condition requires sine wave signals to store energy in the coupling elements. For a coupling capacitor with a long time constant the output transient must last throughout this charging time. When the signal is turned off, some energy is left on each coupling capacitor, which must be returned to the circuit. When the signal is turned off, the output voltage must follow this discharging process.

A good ac design exhibits an optimum recovery process. Time constants are not made any larger than necessary, and the number of coupling elements is kept to a minimum. If several coupling elements are required, it is good practice to make the first time constant as short as possible.

When equipment is turned on or off, coupling elements that require charging can produce an unwanted output lasting seconds. In some cases these transient signals can damage equipment or produce undesirable side effects. One way to avoid this problem is to design the device so that the power for output stages comes up slowly and collapses quickly. This limits the output excursion during any transient period.

7.10 STRAIN-GAUGE ISSUES

A strain gauge is usually associated with a resistive Wheatstone bridge. Arms of the bridge in an instrument are called completion arms. Active arms are attached to a structure to sense small dimensional changes. These dimensional changes alter the resistance proportional to stress, strain, or torsion. Pairs of active arms can be arranged to respond to one strain and ignore another. Bonding the gauges to structures to measure mechanical parameters is the responsibility of a mechanical engineer. A load cell is a specific strain-gauge configuration used to measure mechanical loads. These cells can be used in such diverse applications as measuring jet engine thrust or postal weighing machines.

A strain-gauge bridge requires either voltage or current excitation applied across two opposite terminals of the bridge. The other two corners of the bridge supply the signal of interest. The signals that result are often in the millivolt range, requiring high gain and careful shielding. To ensure that the excitation voltage at the bridge is correct, sense leads are often brought out to the gauge. This avoids the problem of voltage drop in long leads. The leads are often fine wire to reduce cost and bulk. The lead resistance can vary several ohms during the heat of a day, which can change the excitation level and thus the signal sensitivity by several percent unless remote sensing is used.

One form of calibration shunts a known resistor across an active gauge element. This change in resistance corresponds to a known stress in the gauge element. To avoid voltage drops in the excitation leads, separate calibration conductors can be brought out to the external bridge. The lead count with excitation, voltage sensing, signal, calibration, and shielding can be as high as

11. For many short-run installations five leads can suffice. Shunting resistors are usually located within the instrument along with associated relays.

Calibration can take the form of one or two shunt resistors that may be connected across different arms of the bridge. If two or more calibration points are required, the switching schemes can be complex. In some testing facilities these calibration resistors are unique to each bridge and are supplied on a plug-in card located within the instrument.

A bridge configuration can be used to measure temperature, pressure, position, and acceleration as well as strain. If the excitation source is balanced (has a midpoint), the bridge can be completed with only two elements. The midpoint of the excitation supply then becomes one of the signal leads. The advantage of this arrangement is that small voltages inserted in series with the signal can be used to zero the bridge or superpose a calibration signal.

7.11 EXCITATION SUPPLIES

A separate transformer and power supply for each gauge is the best way to supply excitation power. Common supplies require a common ground for all gauges. This common ground implies that the gauges cannot be grounded at the structure. If one gauge is grounded, potential gradients along the structure are apt to couple noise into the signal path of the other gauges. If the one power supply fails, the test is lost. In many tests the device under test or the cost of the test is higher than the cost of the instrumentation.

Shielding in an excitation supply transformer deserves some attention. The differential amplifier is isolated by a driven guard system and high input impedances. The excitation circuitry must be at this same guard potential. It is good practice to provide a primary shield and a guard shield in this transformer. Remember it is desirable that leakage capacitance out of the guard be held to 2 pF. Shielding limits the flow of power currents in the excitation leads. If the excitation supply is used floating with one of the signal leads grounded, shielding requirements for the excitation supply transformer become even more difficult to meet. Power current can now flow in the gauge arms and must be limited to a few nanoamperes.

Excitation voltage requirements can vary from a few volts to 30 V. The higher voltages are often used for high-impedance piezoresistive transducers. The voltage limits are usually imposed by the power that can be handled by the transducer. If the power is pulsed, then higher voltages can be used. This also has the advantage of providing a larger signal and an improved signal-to-noise ratio. If a drift error is 1%, a $5:1$ pulse system could reduce this error to 0.2%. Pulsing obviously limits signal bandwidth.

The upper limit of excitation current is usually 100 mA. The excitation must not overheat the gauge, or the measurement may be flawed. The supply should have current limiting to protect it and the transducer. Power supplies with constant-current features can have the current limit set to, say, 110 mA. A typical 100-ohm gauge excited at 10 V dissipates $\frac{1}{4}$ W per gauge element.

A balanced constant-voltage power supply is straightforward. It is simply two power supplies in series. A balanced constant-current source cannot be built from series current sources. One approach is to place a constant-current supply in series with a constant-voltage supply. The voltage source output regulates to the voltage output of the current source. The transducer still sees a balanced high-impedance current source. If the current amplifier has both current and voltage modes, the one design can provide constant voltage, constant current, and current limiting. The feedback processes can be arranged so that the voltage regulator dominates if the current limit is not met. If the current limit is reached before the voltage limit, the current limit prevails.

A constant-current supply should not have a large shunting capacitor across its output. This is counterproductive because the intent is to provide a high-impedance source even at high frequencies. The capacitor also has the effect of limiting the response time when the current is reduced. If the capacitor supplies unwanted current, the best the amplifier can do is turn off. This is a slew-rate situation that needs to be avoided. Typical output impedances are above 1 megohm shunted by 200 pF. Rapid transitions can be required for several volts. The full-range compliance voltage can be as low as 0.5 V and up to 30 V.

In the voltage mode a large output shunting capacitor is desirable. This capacitor can be added by using a FET switch when the operating mode changes. This approach makes it much easier to stabilize the design in both modes. Stability for all loads and all settings must be carefully verified. Typical output impedances are 0.02 ohm in series with 10 μH. The stability issue is discussed in Section 5.14.

One approach to mode switching is to have two series pass elements. It should be possible to turn these pass elements on independently. When turned fully on, these devices look like a short circuit. In the voltage mode the first element can be an emitter follower, and the second element can be set to be a short. In the constant-current mode the second pass element becomes a collector source, and the emitter follower can be set to be a short. It takes far less feedback to lower the impedance of an emitter follower than to lower the impedance of a collector source. Similarly it takes far less feedback to raise the impedance of a collector source than to raise the impedance of an emitter follower. Limiting the open-loop gain requirements makes it easier to stabilize the circuit.

Stabilization of a dual-mode supply in the crossover region can be quite difficult. One solution is to provide some hysteresis to eliminate the transition region. This can be handled by decreasing the current level by a known amount when the current-mode level is reached. This forces the supply to be in one mode or the other and avoids a transition region.

7.12 REMOTE SENSING

It is possible to regulate the static voltage delivered to a remote point. It is not practical to regulate the voltage at high frequencies at a remote point because of transmission time. This delay would make any feedback system with reasonable bandwidth unstable. Hence, it is standard practice to provide dynamic regulation of the power supply at its output terminals.

Remote sensing requires two sense leads to the load. The signal returned must be differentially buffered (forward referenced) to the excitation supply common. The voltage drop in the excitation leads is a common-mode signal. It is good practice to limit the current drawn in the sense leads. If the feedback resistors in the buffer are 1 megohm, the sense current is only 15 μA when the supply is set to 15 V. If the sense leads have a resistance of 10 ohms, the error caused by the *IR* drop is only 150 μV out of 15 V, quite negligible. Remote sensing is not meaningful in the constant-current mode.

7.13 BRIDGE BALANCE

It is standard practice to balance a strain-gauge bridge prior to a test. The zeroing process also eliminates any dc amplifier errors and thermocouple effects. This is not practical if the true zero occurs after the test is started. If the test requires the strain gauge to be fully stressed, the unstressed zero point may actually be off scale.

Balancing (zeroing) is often handled by current insertion into one of the input leads. This current can come from a slider on a balancing potentiometer or from the output of a DAC. In any case this method shunts the elements of the bridge and modifies the very sensitivity provided by the manufacturer. This error is minor if the insertion resistor is large compared to the resistance of the active element.

Another balancing scheme inserts a small voltage in series with the signal at the input lead. This method does not load the bridge and is easy to apply if the excitation source is balanced. This balancing voltage can also be supplied from a DAC.

Automatic balancing is preferred in large systems because manual balancing can literally take days. Upon the command to balance, each channel proceeds through an algorithm and sets a DAC to the correct balancing voltage. The standard technique is to use a successive approximation method. If balancing does not occur, an alarm can be set to indicate a channel needs some attention. Balancing requires some signal filtering at the comparator input, or the scheme will fail. Comparators which check for zero crossings are in themselves high-gain devices. At high gain the amplifier noise when added to the comparator noise causes a false balance to occur. It is practical to balance an instrument to within a few millivolts for instrument gains up to 10,000.

A balancing potentiometer placed across an excitation current source reduces the output impedance to the value of the potentiometer. In most specifications this fact is ignored. Output impedance tests must be made with this circuit disconnected.

In a computer-operated system it is practical to limit automatic balancing to specific channels. This allows established balance settings to be used. After a power shutdown, computer memory (local or remote) can be used to supply DACs with the prior settings. The computer can then verify whether the channels have drifted from their previous balance points.

High-speed balancing can be achieved by noting the offset of the amplifier. If the amplifier is in overload, the algorithm must reduce the gain until the signal is in range. The proper DAC setting can be calculated in software and applied to zero the circuit. This form of balancing can reduce balancing time to a few milliseconds. A quality A/D converter is required.

7.14 SHUNT CALIBRATION ACCURACY

The accuracy requirements of the shunting calibration resistor relate to the accuracy of the full-scale signal. If the gain accuracy is 0.1%, the shunting resistor should be somewhat more accurate. The bridge output voltage for a single active arm is approximately

$$\left(\frac{\Delta R}{4R}\right) V \tag{7.1}$$

where V is the excitation voltage and ΔR is the change in bridge arm resistance.

Shunt calibration is an attempt to duplicate a known strain in a bridge arm. A single active arm is actually nonlinear in R. For example the bridge signal has an error of -1% when $\Delta R/R = 0.04$. When $\Delta R/R = 0.004$, the linearity error is -0.1%. The error signal is approximately

$$-\left(\frac{\Delta R}{4R}\right)^2 V \tag{7.2}$$

In general, a strain gauge with two active arms is far more linear in R than a single active arm. The first error term is third order rather than second order, as given by Equation (7.2). The output signal is double that given by Equation (7.1).

Constant-current excitation for most two or four active arm bridge configurations provides an output signal that is exactly linear in ΔR. The exception occurs when two opposite arms are active. Then the error term is given by Equation (7.2). Note that the output signal is again twice that given by Equation (7.1).

7.15 PIEZORESISTIVE BRIDGES

When a bridge is formed from piezoresistive elements, the signals produced by the bridge are larger. This implies that a reduced gain can be used. The gauge elements are temperature sensitive, and thus balance is not easily maintained. For short tests this may not be important. When dc instruments are used, low-frequency phase shift does not occur. The amplitude and phase relationships might provide all the data needed for a structural modal analysis. This means that drift is not important as long as the instruments do not saturate or clip the signal.

Various long-term balancing schemes can be applied to piezoresistive amplifiers. Any scheme that forces the output to be zero over some extended time span makes the system ac since it does not respond to dc. If the phase shift is sufficiently small at the lowest frequency of interest, this instrumentation approach is acceptable.

7.16 SERIES CALIBRATION

Series calibration involves using the strain-gauge bridge as a resistance. If this resistance is intact, the calibration level is known. This technique can be used where the bridge cannot be balanced before or after a test. A large resistor is placed in series with the excitation source. The amplifier senses the voltage drop across one gauge element. This circuit configuration must be set up with relays in the instrument. The resulting signal departs from normal if any gauge element opens or shorts.

7.17 AMPLIFIER ZEROING

In strain-gauge or low-level instrumentation work, it is common practice to verify the zero of each instrument before a test. A relay or switch in the instrument disconnects the transducer and shorts the three input terminals together. The instrument should be within certain zero limits, or it must be replaced or repaired. In some instruments a floating source is unacceptable. If this is the case, then the switching must ground the input signal before a measure of zero can be made.

7.18 STABILITY AND INSTABILITY

All systems that use feedback in any form might be unstable. Any testing for instability should include various loads, gain settings, and offsets. Tests in manufacture or acceptance need not be as extensive as tests in design. If instability is suspected, the problem should be resolved by careful testing.

Wiring and proximity can sometimes couple output signals into input lines, and this is feedback. Common power supplies can also be a source of feedback. System instabilities are often not present until most of the system is in place. Tests of individual instruments may not indicate a problem, even though the system problem may require changes to the individual instruments.

Instability implies an oscillation or a condition of excessive ringing. Some instability can occur in internal stages and not appear in the output because of limited bandwidth. If the oscillation frequency is high enough, slew-rate processes may limit the amplitude to millivolts. Sometimes instabilities result in offsets or gain errors that cannot be explained. Sometimes instabilities are a function of operating level and may not appear until some unusual operating condition is experienced.

Instability can be a function of operating temperature. At low temperatures, for example, open-loop gain can increase. If there is insufficient margin in the design, an oscillator can result. Instability can sometimes appear in noise patterns. If the noise seems to have content at one certain frequency, this could be a sign of trouble.

Parallel transistors or FETs can sometimes oscillate above 100 MHz. In limited-bandwidth applications these instabilities may go unnoticed. In some cases the instability limits performance or overheats the devices. A safe practice in design is to add small resistors in series with gates, emitters, or bases to avoid paralleling any two elements.

7.19 STABILITY TESTING

A common method of stability testing involves the use of square waves. The response to a square-wave input is a good indicator of any potential problem. The ideal response is a slight overshoot without ringing. It is necessary to check each feedback section separately because the response of an output stage might obscure the instability of a preceding gain block. The presence of a probe internal to a circuit can load that point, causing an instability. The safest approach is to use an oscilloscope probe with a 10:1 attenuator to limit any reactive loading to a few picofarads.

Most stabilized feedback systems approximate the behavior of a second-order system having a complex pole pair of transmission. A second-order system is characterized by its natural frequency and damping factor. A system with a damping factor of 0.7 responds with a 7% overshoot to a step function input. More complex systems with active filters may be quite stable yet have overshoot in excess of 20%. These systems may still have a relatively flat amplitude response with frequency.

Stability should always be checked for both small- and large-signal levels. A small-signal level implies linear operation where the rise time for a step input is not a function of amplitude.

Output stability should be checked for a wide range of capacitive loads. For example, tests might be made at 100 pF, 0.001 μF, 0.01 μF, and 0.1 μF. Excessive ringing or ringing on the leading edge are indications of instability. It is common practice in most designs to place a parallel LR circuit in series with the output. This keeps the amplifier from seeing the load directly. The inductor can be 10 μH and the resistor 10 ohms. The dc R of the inductor can be held to 0.1 ohm.

In some cases a stability test must be made by driving the output stage from the output terminals. In this case a series resistor is used between the square-wave generator and the DUT. An oscilloscope observes the output directly. This test demands current from the output in a step manner. The output signal should be a series of spikes that represent the transition period when the demand for current is changed. The spike should have a slight undershoot without ringing. If the output impedance is low, the output should return to zero after each transition. This test should be made for various offsets and for various reactive loads. A capacitor can be used in series with the generator to avoid loading the amplifier at dc. This test can be used when it is impractical to insert an input signal, such as in a regulated power supply.

Power supplies are active feedback circuits and must be tested for stability. A power supply is a power amplifier with a dc input signal. The fact that a power supply has a large capacitor across its output terminals does not mean the circuit cannot oscillate. The capacitor may limit the amplitude of the oscillation while the circuit appears to be functional.

7.20   STABILITY OF CONSTANT-CURRENT SOURCES

A constant-current source has ideally an infinite source impedance. This impedance is a function of the feedback factor, which falls off with frequency. This means that a typical high-impedance source looks like a high resistance shunted by a capacitance. Typical current sources might look like 3 megohms shunted by 100 pF. Any load placed on the supply during a test will reduce the output impedance, and this includes an oscilloscope probe.

A current source will not function without a loading resistor. The resistor must be selected so that the output voltage is within its compliance limits. For example, a 20-mA setting can supply a load of 1000 ohms if the output can swing 20 V. A test of output impedance involves placing a voltage source in series with a load resistor and noting any change to the current. When the voltage source is 1 V, the voltage out of the current source must also change 1 V. If the current is monitored directly, it should not change appreciably.

A test using a battery for a voltage source measures the output impedance at dc. If the voltage changes 1.5 V and the current changes 0.01 mA, the output impedance is

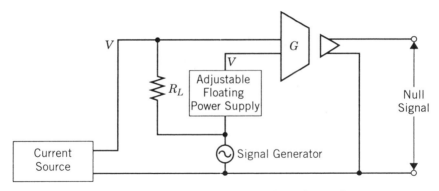

Figure 7.5 A test circuit for measuring the output impedance of a current source.

$$Z = \frac{\Delta E}{\Delta I} \tag{7.3}$$

or 150,000 ohms. This method is not nearly sensitive enough to measure impedances in megohms.

A null method can be used to measure a high output impedance. This circuit is shown in Figure 7.5. A floating adjustable voltage source is placed in series with a signal generator. This voltage is adjusted to equal the voltage out of the DUT. A differential amplifier observes any voltage difference as the signal generator changes level. This amplifier must be capable of handling the input common-mode signal. Because of this null arrangement, the input impedance of the differential amplifier is not critical. The output impedance of the DUT is again given by Equation (7.3), where E is the voltage output of the generator, and the current change is the voltage difference sensed by the amplifier divided by the load resistance. If the generator output changes 5 V and the difference voltage is 10 mV, the change in current is 10 μA. The output impedance is thus 2.5 megohms.

It is possible to use sinusoids to test for the shunt output capacitance. If the 10-mV signal in the foregoing test increases by 3 dB at 1 kHz, the capacitive reactance must be 2.5 megohms at this frequency. This represents a capacitance of 63.66 pF.

In a typical strain-gauge application a high-impedance source is limited by cable capacitance. In a typical cable run the capacitance can be 30 pF/ft. In a 100-ft length the capacitance shunting the source would be 3000 pF. This represents a source reactance at 10 kHz of 5300 ohms, certainly not 2.5 megohms.

When square waves are used to test for output impedance, the nature of the null signal indicates design stability. Ideally, the null response should be a square wave with limited overshoot and ringing. If the null pattern changes amplitude for low-frequency square waves, the design should be challenged.

CHAPTER 8

POWER DISTRIBUTION AND INTERFERENCE

8.1 INTRODUCTION

In Chapter 1 the nature of power distribution and power grounding was discussed in relation to the *National Electrical Code*. This chapter continues the discussion using the basic ideas presented after Chapter 1. Power, when combined with topics such as fields, spectrum analysis, transmission lines, and electric shielding, makes it possible to see how interference must be handled.

Power conductors are usually viewed as carrying the power. It is far more illuminating to accept the idea that power is carried in the fields between the conductors and that the conductors serve to direct where these fields are to go. The power conductors are transmission lines with all sorts of stubs, mismatches, reflections, and other miscellaneous anomalies. Interference of all types uses these same conductors for transport into and out of equipment. It is these phenomena that must be controlled.

Fields exist inside the copper at power frequencies. A small E field parallel to the conductors directs a component of Poynting's vector into the conductors. This is the I^2R loss in the copper. The majority of the field directs the power to flow along the path between the conductors.

8.2 THE ISOLATION TRANSFORMER

The word *isolation* means many things. The user hopes that when an isolation transformer is used, it will solve his or her present or future problem, whatever it may be. Isolation from difficulty is not achieved by guesswork. Certain interference mechanisms can be handled by an isolation transformer, but the transformer must be applied correctly.

Generally speaking, an isolation transformer has come to mean a single-phase transformer with multiple electrostatic shields. These shields are brought out for connections to the user's system. A manual describing the use of these shields is usually not provided. The simple approach is often to "ground" the shields to a "good ground" and hope for the best. If this does not work, then the "ground" is not good enough. Such reasoning is all too common and confuses the true issues of controlling the transport of interference.

The discussion starting in Section 3.3 shows how shields are used to direct current flow to reduce common-impedance coupling. At low frequencies, where circuit ideas can be used, it is convenient to consider the current path rather than the location of fields. Shields are connected to direct the current flow away from signal conductors. Any interfering current flowing in a signal conductor is sensed as signal and amplified. This treatment of shields in Chapter 1 was intended for application within electronic equipment, not for shielding in the power distribution system. Of course, an isolation transformer could be mounted inside a rack of hardware to supply power. In this case the NEC may have no jurisdiction, but the question must still be asked as to exactly what can be expected from this solution.

Within an instrument there can be up to three shields. This solution is expensive and not often used. The preferred solutions involve high-input-impedance differential amplifiers. The primary shield connects to the equipment ground, the only power-related connection. The other shields are signal common or input guard, and these conductors are not shared between devices. This means that an isolation transformer supplying power to several devices can supply only a common primary shield, since there is no obvious way to connect the remaining shields to various signal grounds without modifying the electronics in some strange way.

There is a way to use three shields in a power transformer. Such shielding is covered in Section 8.6 when separately derived transformers are discussed. These transformers do provide a form of isolation, but it is not industry practice to call these devices isolation transformers.

The shields in typical isolation transformers have mutual capacitances below 1 pF and often below 0.1 pF. This means that the shield forms a wrap around the coil much like a gift box might be wrapped. In a test for these capacitances, the test wiring must be properly shielded (see Section 3.3). To avoid violating this capacitance, this same shielding must be extended over all conductors connected to that transformer coil. This may require shielding primary conductors, for example.

Isolation transformers installed such that the primary and secondary conductors share the same receptacle or conduit violate the shielding provided. A simple transformer without shields will work just as well and will cost far less. When a transformer is mounted outside of equipment, primary and secondary conduit must be bonded together. These conductors are equipment grounds and must return fault current back to the nearest line interrupter. Any attempt to float the conduit is unsafe and violates the NEC.

Isolation transformers have been built with internal electronics. This electronics smoothes the waveform so that short gaps in power or excessive spikes are removed. In this sense the transformer isolates the user from a noisy power source. This is certainly a different meaning of *isolation.* A group of shields cannot provide this performance. A power transformer is supposed to couple voltage waveforms exactly. There is no way that a shield can allow the transfer of some voltages but not others.

A power distribution transformer used as a separately derived system does provide a form of isolation (see Section 1.10). This added transformer provides the user with a new neutral conductor and a short fault path for equipment grounding. This new neutral is grounded to the nearest point on the grounding electrode system. This type of installation makes it possible to control common-mode interference arriving on the power conductors. Using an isolation transformer this way implies that it is a part of the facility wiring and that it is rated and listed for this type of service.

A transformer used to buffer a source of power is a form of passive filter. The leakage inductance and coil capacitances form a T-type low-pass passive filter. This filtering action can sometimes be of benefit. The presence of shields may actually help to increase the leakage inductance. This filtering action is not repeatable between transformers, and in some cases resonances may complicate the problem. When filtering is needed, it is preferable to use a known repeatable component rather than a transformer.

8.3   THE TWO TOOLS OF INTERFERENCE CONTROL

The two tools available to control interference are the metal box and the ground plane. The box idea was developed in Section 2.5. Electrical activity inside an ideal metal box cannot leave the box, and activity outside the box cannot get in. In the world of real circuits, the metal box must be penetrated by power, signal leads, control leads, display devices, and ventilation holes. Seams and access must be provided so that the hardware can be built. The interference game that must be played is to plug these holes and seams and filter leads that bring in interference.

The second tool, more subtle, is a plane of conductive material. The concept of ohms per square was discussed in Section 4.7. A large conducting surface is often called a ground plane, although the use of the terms *ground* and *plane* can be misleading. The metal skin of an aircraft is a ground plane, yet the aircraft need not be earthed. The word *ground* adds electrical meaning to the word *plane,* which also means aircraft. In addition, a ground plane need not be flat to be effective. Such is the semantic maze we live in.

A plane wave cannot penetrate an infinite conducting surface because the E field tangential to the surface must be near zero. The wave impedance is

377 ohms (ohms per square), and the conductive surface may be around 100 microhms per square. This mismatch in impedances means that an impinging plane wave will be almost totally reflected. Any energy that does enter the surface will be absorbed (skin effect).

A conductive surface will support a plane wave moving along its surface if it has a vertical component of the E field. A plane wave with a horizontal component of E cannot exist near the surface because the surface current would have to be enormous. The lack of any horizontal E field component is the dominant feature of a ground plane. If this feature is not used, then the ground plane is not used. Note that the ground plane provides this characteristic whether it is earthed or not. A ground plane is grounded (power meaning) for different reasons.

Section 4.17 develops the concept of E field coupling into loops. If the return conductor in Figure 4.4 is a ground plane, the E field can have only a vertical component. The coupling is proportional to loop area. If the signal conductor is located on the ground plane surface (insulated), the coupling is essentially zero. The lesson is simple. A ground plane is useful only when wiring and cabling rest on the ground plane. Merely having a ground plane is not sufficient. As with any other tool, if it is not used correctly it will not work.

8.4   POWER LINE COMMON-MODE SIGNALS

Power conductors travel parallel to the earth over most of their run. The loop area formed between the conductors and earth couples common-mode signals to the power conductors. This field propagates along the power conductors in both directions and couples into every piece of hardware using power. The signals of concern here generally have frequency content well above the power fundamental or its harmonics. In a circuit sense, any coupled common-mode voltage circulates current through the mutual capacitances of the transformer into the secondary circuits. The return paths include the secondary terminations to various grounding points. This current flow in secondary commons causes difficulty. Any current flowing in a circuit common adds interference to the signals of interest. This form of interference is correctly called common-impedance coupling. In some installations this type of interference can cause circuit damage. Typical common-mode signals result from power switching, lightning, and various radio and television transmitters.

The best way to limit common-mode current flow in secondary circuits is to provide a low-impedance shunt path and raise the impedance in the series path to interfering signals. These solutions involve passive filters or transformer shielding. The shunt path can be a filter capacitor, and the series path the mutual capacitance of a shield. Unfortunately, a low-impedance shunt path is not always available. If the path is over a single equipment grounding conductor, it can be highly inductive. This inductance makes any shunting capacitor ineffective and limits the benefit of the shield. The inductance allows some of the common-mode current to flow in the secondary circuits. A better filter or shield will not improve this situation. The filter problem is shown in Figure 8.1. The most severe problem occurs when a dedicated equipment grounding conductor is used. (Isolated grounds are discussed in Section 1.9.)

A shielding solution exists if there is a primary shield in each piece of equipment. This shield is usually not provided by equipment manufacturers unless required in specifications. Even if shields are provided, any impedance in series with the shields reduces their effectivity. As discussed earlier, an isolation transformer with a single shield can be used to power many items of hardware. This one transformer thus becomes the buffer for limiting common-mode current flow. Again, if the equipment grounding path is inductive, the shield's effectivity is reduced.

An added shielded transformer (distribution or isolation) can be made effective with respect to a local ground plane. This is also true for a line filter. The impedance path to the ground plane can be kept quite low. If the lead length is inches, the inductance can be as low as a few nanohenries. Now if the secondaries are referenced to this ground plane, very little power line common-mode current will flow in the secondary commons. The transformer shield solution is shown in Figure 8.2. It is easy to see that there is no best ground or better ground. Any ground is ineffective if it is located at a distance. The only way to attack this problem is by associating the hardware and cabling with a local reference conductor (ground plane).

The primary coil of the transfomer in Figure 8.2 can be considered to be in "power space" or in the power environment. The secondary voltages can be thought of as being in the "signal space" or ground plane environment. Common-mode field energy is reflected at the transformer interface and not allowed to propagate onto the ground plane. The chief function of the ground plane is to provide a method of limiting the flow of common-mode current

Common-mode current flows in C through common-impedance R. The inductance L in the equipment ground path does not allow the filter to be effective.

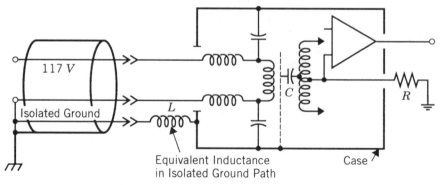

Figure 8.1   Common-impedance coupling and line filters.

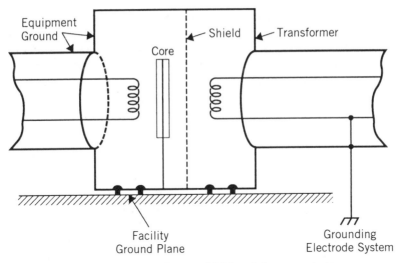

Figure 8.2 The transformer shield and the ground plane.

in signal conductors. When current does flow in the ground plane, it produces no significant tangential E field and thus does not contribute to signal coupling.

8.5 DIFFERENTIAL POWER LINE INTERFERENCE

Electronic loads are often highly nonlinear. Current is drawn at peak voltage or in step manner at various points along the cycle. These current demands imply large H fields at harmonics of the power frequency. All power lines are somewhat inductive, and, in a circuit sense, nonlinear loads modify the voltage waveform. These changes to line voltage are differential in nature and by transformer action will appear on the primary and secondary coils of any subsequent transformers. Shields cannot eliminate this phenomenon.

Differential-mode distortions of the power voltage waveform are usually a maximum at or near loads and appear to be less at the service entrance, obviously the result of voltage drops in the line impedances. If dedicated branch circuits are used for critical loads, distortions in waveform can sometimes be avoided. Voltage waveforms can be improved through the use of differential line filters. These filters must handle harmonics of the fundamental and can be physically large and expensive. If these filters have capacitors that terminate on equipment ground, then any common-mode content is also filtered. This type of filter can result in significant equipment ground current.

Three-phase systems that handle nonlinear (electronic) loads are apt to have large neutral current flow. For a balanced linear load the fundamental frequency neutral currents cancel. When there is high harmonic content, cancellation does not take place, with the result that the neutral conductor often overheats. A neutral conductor with a significant voltage drop adds noise to a system. The neutral voltage drop appears between neutral and equipment ground. It is common-mode in nature and circulates current in secondary commons unless there is shielding or filtering. A wye/delta transformer in the power path allows a redefinition (regrounding) of the new neutral. The neutral conductor on the new secondary will not carry neutral current from the previous power section.

8.6   COMMON-MODE REJECTION IN POWER SYSTEMS

Filtering the power line at harmonics of the power frequency is difficult and expensive. It is usually cost effective to use different branch circuits or add a separately derived system to limit problems caused by nonlinear loads.

Transformers with shields are the best way to divert and control common-mode current below 100 kHz. At higher frequencies the mutual capacitances begin to dominate the performance. Above 100 kHz the passive filter provides a good solution to controlling interference, provided there is a reference conductor (ground plane) available.

Shields in power transformers have mutual capacitances (leakage capacitances) that often exceed 25 pF. This corresponds to a reactance of around 6000 ohms at 1 MHz and a current of 10 mA for a spike of 60 V. In many circuits with a few microhenries of lead inductance, this current can generate a voltage pulse of 0.1 V. Shields are not effective at these frequencies, and a combination of shield and passive filtering is often needed. Shields are effective for power phenomena, and filters are effective at frequencies where the shields are ineffective.

The single shield in a distribution transformer can be effective in controlling common-mode effects at power-related frequencies. The shield is internal to the transformer and is grounded, ideally on the reference or ground plane, where the transformer is mounted. By Code, the neutral on the secondary must be grounded to the nearest point on the grounding electrode system for the facility.

8.7   CONVERSION FROM COMMON MODE TO DIFFERENTIAL MODE

Common-mode current flows from the primary coils of the transformer to the primary shield. Some of this current flows in turns of the primary coil and is converted by transformer action to differential interference on the secondary. This current level is obviously frequency related. The voltage appearing on the secondary depends on the load impedance.

To avoid this conversion process, a second shield can be introduced in the transformer. This shield is shown in Figure 8.3. The shield is connected to the grounded conductor side of the primary. Common-mode current now flows between the shields and not in turns of the primary.

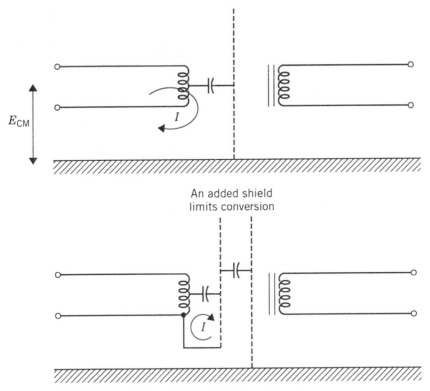

Figure 8.3   A second shield in a power distribution transformer.

This type of internal shield can also be added to the secondary winding. If a nonlinear load modifies the secondary voltage, part of this voltage disturbance is common mode in nature. An added shield can reduce conversion from common-mode to differential-mode coupling back to the primary.

Commercially available power centers provide this type of distribution transformer together with passive filtering. The NEC allows listed equipment of this type to have the secondary neutral grounded on a ground plane rather than to a point on the grounding electrode system. These centers often provide circuit breakers, meters, transient protection, and matching cabinetry. They are ideal in systems that require immunity from most of the standard power distribution problems. In this form they do not provide voltage regulation or uninterrupted power.

8.8   GROUND PLANES

The ideal ground plane is a solid sheet of metal extending under all equipment and wiring. This plane can be as thin as 1 mm of copper; however, this thickness is not robust enough to be practical. A practical solution might be sheets of 0.02-in. cold-rolled steel welded together. Another solution might be strips of copper 4 in. wide bonded together to form a grid on 18-in. centers. The grid is not quite as effective as a solid sheet of metal because currents must flow around bends to go between two points. This form of grid can be effective for frequencies below 20 MHz.

Computer floors made up of a grid of stringers are often considered ground planes. A solid conducting surface has an ohms-per-square measure below 100 μohms over a wide frequency range. To be effective the stringer bonds must be at least this good. To meet this standard the connections between stringers requires plated surfaces and pressure-type washers. The bond should be sealed so that the contact area will not corrode in time. Equipment should not be located out on the edge of this grid where the ground plane is not effective.

Floor tiles are used to provide a walking area and form the plenum chamber below. These squares should have some conductance to bleed off any charge accumulated on a person walking on the floor. This impedance should be around 10^9 ohms per square, giving a person with a capacitance of 300 pF a 0.1-s time constant. A continuous connection must be made between the tile and the stringer system if the tiles are to be effective. To limit the threat of electrostatic discharge (ESD), these floors should never be cleaned with a rotating mechanical brush.

In most installations a raised-floor ground plane is not effectively used. Cables that connect between cabinets are draped on the floor under the ground plane. The loop areas formed by this pattern allow common-mode signals to couple to the loops. Rarely are the racks bonded to the stringers to extend the ground plane into the equipment. Ground straps, if used, are not very wide and may not connect near a cable exit. Connections are often made by equipment grounds or signal grounds involving a large loop area. It is a better idea to place the ground plane on the floor under the raised stringers and get some use out of it. Cables draped on the floor are now on the ground plane, eliminating loop areas. This approach still requires an extension of the ground plane into each rack. This can be handled with a wide conducting strip bonded to the ground plane and to the rack across its entire width. Cables leaving the rack must follow and lie on this conductor. This approach is shown in Figure 8.4.

A philosophy often exists that requires a ground plane somewhere, particularly in a big system. This is the case just discussed, where cables are routed several feet from the ground plane and bonding to racks is minimal. This approach is accepted as long as there are no problems. It can be argued that because these systems work, the ground plane layout is correct. It probably means that equipment manufacturers are able to design equipment that performs well in the presence of common-mode signals.

A ground plane consisting of buried rebars (reinforcing steel bars) cannot be very effective. First, the rebars cannot be seen. This means that their extent is uncertain. Second, new connections to the steel are difficult to make, and

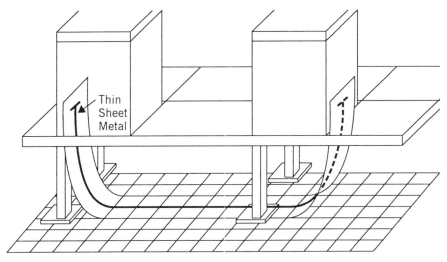

Figure 8.4 A proper ground plane.

once made are difficult to move. Third, there is no guarantee that the rebars are properly bonded at each intersection to form a grid. Bonding weakens the steel unless each weld is carefully done. This treatment is expensive.

A question often asked involves extending ground planes into adjacent rooms. Ideally, ground planes should be continuously connected. A compromise solution is to connect the ground planes together through the wall every 6 in. with, say, No. 10 wire. Each connection represents an inductance, and larger conductors serve no purpose. Round conductors are easier to handle because small holes can be drilled in the wall.

When a ground plane is extended by a tray or a narrow run the impedance is calculated by noting the number of squares that make up the run. A tray 6 in. wide and 10 ft long represents 20 squares. If the ohms per square is 100 μohms, the impedance of the extension is 2 milliohms, assuming a bond between the last squares and the two ground planes. A tray constructed of sheet metal screws is not acceptable. A layer of copper on the bottom of the tray bonded at both ends is acceptable.

When ground planes must be extended to two floors of a building, the ground plane must connect by using one of the walls. This requires planning during initial stages of construction, since it can be quite difficult after construction.

Extending a ground plane to serve two buildings may not be impractical. The earth itself is a ground plane, but any cable connection involves an undefined loop area. In wet weather the loop area is much smaller than in dry weather. It is better to plan communications that rely on fiber optics rather than solve the common-mode problem through the use of ground planes. Fiber optics are not sensitive to lightning, but an electrical connection can pose difficult problems.

8.9   GROUNDING GROUND PLANES

A ground plane is part of the grounding electrode system of a building and therefore must be bonded to all other building conductors. The bonding must protect equipment when lightning strikes. Consider a 10-story steel structure. When lightning strikes, the current is seeking earth. The inductance of the down conductors can be 1 μH/ft. If the current uses one of the steel elements and the height is 120 ft, the inductance is 240 μH. If the frequency of interest is 640 kHz and the current is 10,000 A, the voltage drop is 9.6 million V. On the fifth floor there can be a voltage difference of 1,000,000 V between the steel on opposite sides of the building. If the ground plane is grounded to one point on the building steel, there might be a side flash from the ground plane to the steel on the opposite side of the building. This arcing might use the hardware as its path and destroy it.

To avoid this side arcing to building steel, it is necessary to bond any ground plane to nearby building steel at many points, not to one point. Lightning current flowing uniformly in a ground plane poses little threat. If the ohms per square is 100 μohms, the entire 10,000 A develops a potential drop of only 1 V across the floor. At points of current entry the voltage drops are somewhat higher. In any case the side flashes would not occur, and any equipment on that floor would not be affected.

It is good practice to provide a ground ring around the ground plane in a facility. All conduit entering the area should be bonded to the ground ring. The ground ring is, in turn, grounded to the building steel and the ground plane. This approach keeps any conductor from transporting a high-voltage pulse into the ground plane area. It also means there will be no arcing to these conductors in this area.

Inductors should not be placed between the ground plane and its connections to building steel. During lightning activity these inductors will probably vaporize. If the neutral in a power center bonds to the ground plane, these inductors could limit fault current and, thus, are illegal. A ground plane is no place for single-point grounds: the more metal connected together, the better. The only required single point ground connection is the neutral connection of a separately derived system to the nearest point on the grounding electrode system.

8.10   PENETRATION OF FIELDS INTO BOXES

The small FM radio experiment in Section 4.14 illustrates how difficult it is to keep field energy out of a box. The experiment shows that it takes only one conductor to bring field energy into a metal enclosure. The lesson is obvious. To keep field energy out of a region every lead must be filtered. Any lead may be bonded to the enclosure but on the outside. If the lead forms a loop inside the box and then bonds to the box, the loop will radiate into the box.

A control, such as a switch or potentiometer with a long shaft, can be a source of field penetration. If the component bonds to the inside surface and the shaft extends through a hole to the outside, radiation can enter the box since this is a loop.

Radiation penetrates the conductive surface based on skin effect. For most solid plates of metal this penetration is not significant. Near inductive fields can, however, be a problem. See Section 4.4.

Radiation couples through apertures and waveguides based on fractions of a half-wavelength. If the aperture has depth, then attenuation based on a waveguide beyond cutoff can further attenuate the field.

8.11 FILTERING POWER INTO AN ELECTRONIC ENCLOSURE

Power line filters are often added to equipment. This topic is covered in Section 4.15. Combinations of filters, fuses, and switching are often supplied as one component. These components are often mounted on a metal surface with the wiring connections made on the inside of the box. It may be convenient, but this wiring technique violates the filter. All leads must be filtered before entering the box, not after.

An equipment grounding conductor that enters a box allows radiation to enter the box. A filter cannot be placed in this lead because it violates the NEC. Any impedance in the equipment grounding path will limit fault current. The proper way to terminate the equipment grounding conductor is shown in Figure 4.3. Current flowing on this conductor stays on the outside surface of the box.

8.12 PASSIVE FILTERS FOR NOISY CIRCUITS

Many circuits radiate energy into their own environment. Leads which leave an enclosure can couple to this radiation and then carry this energy into space. The only reference point for filtering is the enclosure. This filtering must be inside the box. A loop of wire after the filter can recouple to the energy, and the filter becomes ineffective.

If the box is not bonded to an external ground plane, there is no way for the filter to function with respect to this external reference conductor. Any common lead or shield brought to the box is just as good a radiator as any other lead. Radiation also occurs if the leads required to operate internal circuits must supply any high-frequency power.

Filters plugged into a wall outlet cannot attenuate common-mode interference at the device because the filter is mounted to the receptacle not to the outside of the equipment case being protected. These filters are often supplied with built-in transient protection in the form of a metal oxide varistor (MOV).

This protection can limit common-mode and differential-mode transients with respect to the equipment ground at the receptacle.

8.13   OPTOISOLATION

It is common practice to optically isolate various control signals from the circuit in question, to avoid an ohmic connection to the control leads. The isolation could be done by relays, which are bulkier and more expensive. The optical isolation discussed here takes place at the base of a transistor. Current flowing in a photodiode controls transistor operation. When current flows, the transistor is turned on and functions like a switch.

It is usual to place this form of optical isolation on a printed circuit board along with all other circuits. Control leads can carry field energy into the equipment, negating in part the isolation afforded by the optoisolators. In severe environments it may be necessary to mount the isolators at the bulkhead with the control leads outside and the transistor connections inside. Control leads are a two-way street. They can also couple and carry fields out of equipment. This bulkhead treatment can keep these leads from transporting internal energy to the outside world.

8.14   FILTERING AND ISOLATED GROUNDS

The energy reflected at a power line filter must return to the power source. This return path involves the equipment ground impedance. For a single piece of equipment this impedance causes no difficulty. Filtering mounted on the bulkhead makes this conductor the local reference point. Filter currents will not flow in signal commons as long as the equipment is self-contained (not grounded).

When pieces of equipment are interconnected and separate equipment ground conductors are used, some of the filtered current flows in the interconnections to find parallel equipment grounding paths. The only way to limit this current flow is to bond all equipment grounds together, in effect forming a ground plane. Avoiding isolated grounds will reduce the impedance in the equipment grounding path, which further reduces the current flow in the signal interconnections. A full ground plane treatment does not reduce the filter current returning in the equipment grounding conductors to the source of power. It simply keeps this current from flowing in signal interconnections for that equipment referenced to the ground plane. When noise is kept out of signal processes, the interference problem is solved.

A good example of the foregoing problem occurs when computers drive several graphics display modules. Each piece of equipment has its own line filter. If isolated grounds are used, some of the filtered current flows in the video coaxial interconnect. These signal paths are not differential, and the current adds noise to the video signal.

Equipment mounted in racks sometimes relies on the racks to be a pseudo ground plane. Its quality depends on the bonds between rack sections. If the performance improves when all the screws are tightened, the facts speak for themselves. Straps should be added to guarantee many good connections. Relying on simple sheet metal bolts is not a good plan.

8.15 OSCILLOSCOPE GROUNDING

Oscilloscopes are basically single-ended devices. The fact that A and B inputs can be used for subtraction does not make them differential in the sense of limiting current flow on all signal leads. On each input probe the input shield carries signal current. If noise current flows in this shield, there can be common-impedance coupling, which produces an unwanted signal.

Input probes vary in cost and performance. A better sheath will have lower common-impedance coupling. An expansive metal front panel is no longer available on today's smaller instruments. This makes it very difficult to add a second shield to the probe or to find a screw to add a bonding strap. Hardware can be added to the input connector to provide for a bonding strap.

Oscilloscopes used in high-field environments can "receive" signals through their case. It is always wise to short the probe at its tip and connect the probe common to the circuit of interest. If a signal is observed, the test or measurement will not be valid. A source of noise is the current flowing in the loop formed by the input probe, oscilloscope frame, and power transformer to the power lines. This problem can be controlled by using a ground plane and bonding a line filter to this plane. The oscilloscope and DUT must also be bonded to this ground plane to limit the coupling loop area.

8.16 POWER FACTOR CORRECTION

In most industrial areas of the world the power is quite reliable and voltage fluctuations are minimal. Industrial loads require large lagging reactive current for motors and generators. The utilities prefer to correct the power factor by adding capacitive loads to the lines near the load. In this way they do not need conductors and equipment to support current flow that carries no real power. The capacitors are required during operating hours and are adjusted or removed when a power factor correction is not needed.

Power factor correction capacitors can be placed line-to-neutral or line-to-line. The line-to-neutral solution requires lower-voltage capacitors, and it is easier to handle switching at ground potential. Unfortunately, this technique adds reactive current flow to the neutral. This current is small at the fundamental frequency but can be much higher when the loads are nonlinear. Power factor correction can be provided by the user to reduce the utility bill; it is then the user's responsibility to switch the capacitors.

When a capacitor is switched on the line, it would be preferable to make the connection at a zero-voltage crossing since this limits any current surge. This is not practical on a three-phase system where zero crossings occur at different times. This type of precision switching is also not easy to provide. It is also possible to limit the surge current by adding a series resistor that is removed shortly after a connection is made. These schemes are an attempt to limit any significant disturbance to the voltage waveform. Any switching disturbance is filtered by the system and is greatest near the point of switching.

8.17 LOAD SWITCHING

When a switch closes, demanding power, the power must be transported from the generator in fields between conductors. At the moment of closure the only energy available is stored in the fields that are present. The voltage drops to a value that depends on the load and source impedances. Often in the first microsecond the voltage drops to near zero. In the next few microseconds field energy from the many parallel paths in the facility begin to fill the demand. Within a millisecond the request has made it onto the high-voltage transmission line where a great deal of field energy is located. The wave front arriving at the generator requesting power is no longer steep. In effect, the transmission system functions as a giant distributed energy reservoir. Within a few milliseconds the new power demand is sensed by the generator.

When power is turned off, the energy in transit cannot be turned off. This excess energy must somehow reflect and dissipate in various loads. If the load is inductive, a high-voltage pulse can be generated at the load. This pulse can arc across the switch contacts and result in a very steep wave front that propagates over the power network. This steep wave front is also filtered and attenuated by the system. Power users near the arcing will probably feel the largest impact.

Arcing occurs when an inductive load is interrupted. The interrupted current begins to flow in any shunt capacitance which forms a tank circuit. All energy stored in the inductance must transfer to the capacitance within one quarter-cycle of the tank circuit's natural frequency. If the capacitance is small, the voltage must be very high. Typically this rise time lasts a few microseconds, hardly time for the switch to fully open. When the voltage reaches the breakdown potential of air, arcing results. Arcing generates frequency content that can extend to hundreds of megahertz, whereas the natural frequencies involved in the high-voltage generation rarely exceed 100 kHz.

Within a facility the various demands for power cause spikes to appear on the power line. Some of these spikes are the result of arcing. They have little impact on power supplies within equipment. They do, however, have common-mode content that can circulate current in secondary circuits. The normal defense is a power line filter to equipment ground or a common-mode shield. If the equipment ground impedance is high, then filters and shields are ineffective. This results in a spike of current flowing in secondary circuits. If the common impedance is high (long lines), these spikes can be significant. Power line transients have been known to destroy circuitry at digital interfaces. Reinstating standard equipment grounds reduces this current to safe levels.

8.18 POWER FILTERS AND REMOTE VANS

If the power line is filtered at the van, it must be bonded to the vehicle frame. Filter currents must return to the source of power over any interconnecting equipment grounding conductor. The impedance of a long cable can cause a potential difference between the neutral or common and the equipment ground. This is a common-mode signal that can circulate current in secondary circuits. The filter, in effect, causes problems that might not exist if the filter were removed. Remember, the neutral *cannot* be grounded to the van.

Even if the van has its own separately derived power transformer, it must still be grounded via the incoming equipment grounding conductor. The secondary neutral must be bonded to the van's frame. Any power line filter on the secondary of the transformer must still bond to the van. Filter currents need only return to the transformer neutral; thus there is no common-mode signal to cause secondary current flow. In this system a line filter bonded to the van frame does not cause a problem.

8.19 ON-LINE POWER SUPPLIES

To avoid a line transformer, power supplies can be designed to draw power directly from the utility line. Separation from the utility ground can be provided by a high-frequency switching transformer that converts line voltages to usable circuit voltages. These secondary circuits are usually grounded by users when the circuits are used. This grounding provides a path for switching currents to flow. These currents can cause radiation and affect the performance of nearby equipment. For this reason these current levels must be controlled.

Utility power can be rectified without a transformer. Rectification can be a full-wave bridge or a symmetrical half-wave circuit symmetrical about the grounded conductor (two diode rectifiers). For a 120-V power line the full-wave bridge provides 168 V dc with the power coupled twice per cycle. In the symmetrical case, the voltage is ± 169 V, with power supplied to each power supply once per cycle. On the positive voltage peak the positive power supply capacitor is charged. On the negative voltage peak the negative power supply capacitor is charged. In either system power is drawn from the utility twice per cycle.

In the full-wave-bridge case the power line is essentially disconnected from the power supply except when the diodes are conducting. In the half-wave center-tap system the power supply common is connected to one of the power line conductors at all times. If the line plug is polarized, this connection would normally be selected to be the grounded conductor (power meaning).

Full-wave bridge rectification on the secondary of a transformer isolates the rectified common from the transformer when the diodes are not conducting. This can modulate any common-mode current flow 120 times a second. The current flows only when the diodes conduct. In sensitive circuits this form of rectification is not recommended. Full-wave center-tapped rectification is preferred because this leaves the load connected to the secondary coil at all times. The cost of a tap is thus added to the transformer. If only one power supply voltage is involved, the windings are not efficiently used, which increases transformer size and cost.

Servicing power supplies that do not use a power transformer can be dangerous. The oscilloscope housing is normally tied to equipment ground. If the probe common is connected to the circuit common and this happens to be the "hot" side of the power line, sparks will fly. To avoid this situation, the equipment ground connection is often lifted. The oscilloscope case can then be connected to the "hot" power conductor. This condition is unsafe and should not be allowed. It is much safer to work with on-line power supplies using a 1 : 1 isolation transformer. Its secondary winding is left floating. Transformer isolation allows the oscilloscope to be grounded at all times. The isolation transformer should be rated well above the power level used to provide a low-impedance source of power. The transformer need not have internal shields to be effective. Tests involving radiation and conductive interference should be made when the circuit is powered directly from the mains. During these tests the oscilloscope should not be connected to the circuit.

8.20 OPERATION OF SWITCHING SUPPLIES

A switching power supply usually converts energy stored on the line side of the circuit to energy stored on the secondary side of a high-frequency transformer. This energy can be stored in an inductor or capacitor. The switching rate is usually above 20 kHz and can be above 100 kHz. Switch elements are FETs or transistors that act as short circuits when turned fully "on" and open circuits when turned fully "off." The selection of switching device depends on drive requirements, saturation characteristics, hold-off voltage, dissipation, and cost.

Switching produces a steep leading edge for the voltage applied to a transformer coil. This leading edge is coupled by capacitances to the secondary common. The resulting current must be blocked by a filter or canceled by a balancing scheme. By transformer action the step voltage on the primary appears on the secondary coils. The secondary circuits can rectify this voltage and store energy in a filter capacitor.

When energy is stored in an inductance, a current is allowed to build up. When the stored energy is sufficient, a switch turns the current off. The voltage across the inductance must reverse in direction because the current must now decrease. Voltage reversal allows a diode to steer the current to a new storage capacitor and load. The diode can be on a secondary coil or on the primary circuit. If the switch is left on for a longer time, then more energy is stored in the inductance. When this energy is transferred to a secondary capacitor, the voltage will be higher. Thus, the duty cycle controls the secondary voltage. A feedback system can be used to control the output voltage.

An inductance that stores energy is usually a gapped device. The field that stores this energy is mainly confined to the gap. The transformer current must also flow in any leakage inductance. Since the current flows in pulses and not over the entire cycle, the peak current must be proportionately higher. This results in a larger near field that may produce interference. Leakage inductance is proportional to the square of the turns, and thus it is desirable to keep the number of turns low. This must be balanced by the magnetizing current requirement, which increases linearly as the turns are reduced.

Integrated circuits are available that control the operation of the magnetics and switching devices and produce regulated voltages. Manufacturers of these devices provide circuit design notes with instructions on how to construct the magnetics. A great deal of work has been done in the last few years to improve the nature of these switching supplies.

Switching circuits that do not store energy in an inductor rely on energy stored in a capacitor, or else they take their energy directly from the power mains. This is not advisable because conductive interference can result. These circuits drive a conventional transformer with square waves. The core is not saturated at any time.

8.21 FILTERING ON-LINE SWITCHING CIRCUITS

When power is taken directly from the line, the frequency content of the current flow must be considered. Conductive radiation (current flow) is a parameter limited by FCC and European standards. When energy is first stored in electrolytic capacitors, the frequency content of power frequency current pulses is usually not a problem. Designs with significant problems involve switching circuits with little, or no, local storage of energy.

Electrolytic capacitors are not efficient in yielding their energy at high frequencies. When a switching regulator asks for current in a step manner, some of the demand may be passed on to the power line. When the power diodes are not conducting, the diode capacitances are in series with the power source and no other power source is available. When the power diodes are not conducting, energy can only be taken out of the storage capacitors. When the diodes conduct, the switching demand can result in significant line current from local storage and the power line. These pulses of high-frequency current can easily exceed limits imposed by regulations. Passive line filters (series inductors and shunt line-to-line capacitors) can be used to reduce the level of this high-frequency current drawn from the power line.

The local energy storage capacitor must be wired to supply high-frequency current, that is, as a four-terminal device. The load currents must be taken right at the capacitor terminals on separate leads. These two leads should ideally be coaxial cable. If circuit trace is used, the traces should have no significant loop area. Wiring to the transformer should also not add loop area. A small decoupling capacitor located at the transformer can supply leading-edge energy near the transformer.

The switching process requires rapidly changing voltages on the coils of the switching transformer. Voltage switching can cause current pulses in the primary-to-secondary capacitances. If the power line is connected to the circuit (half-wave center tap) and the secondary is grounded, this current flows in both circuit commons and returns via the facility grounding. This circuit is shown in Figure 8.5.

If facility grounding is a ground plane, the problems may be insignificant. Without a ground plane the inductance of the loop can be substantial and the fields coupled into the facility can be disruptive.

A switching transformer does not accommodate shields. Any shield adds shunt capacitance and leakage inductance, which limit transformer performance. Because of fast leading edges, even small mutual capacitances circulate significant current. Available tools for limiting common-mode current flow are passive filters and winding symmetry. If the secondary coil is wound so that the capacitances are in series with oppositely moving voltages, the current will first-order cancel. A center-tapped primary coil on the switching transformer is required to obtain this symmetry. A small capacitor can sometimes be added to the circuit to obtain first-order cancellation.

Common-mode filters can be placed between the power line and equipment ground. Often at the frequencies involved, the inductance in the equipment ground path will limit the performance of this filter. Within a single piece of

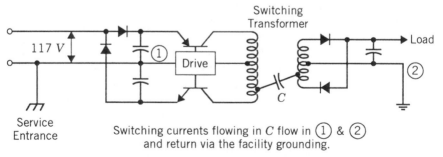

Switching currents flowing in ① flow in ① & ② and return via the facility grounding.

Figure 8.5   The path of switching currents in a facility.

equipment this filtering can be bonded to the equipment housing and the filter can be quite effective. It is still desirable to limit the current flow by winding symmetry and to avoid filters if possible.

If energy is taken from the line with a full-wave bridge rectifier, the line is disconnected from any load except when the diodes conduct. If energy is taken from the storage capacitors when the diodes are open, the circuit is floating except for diode and circuit capacitances. With logic it is possible to interrupt the switching process during power diode conduction. Secondary energy storage must be available to supply power during off time.

The flow of noise currents in the equipment ground conductors can sometimes trip ground fault interrupters. The average current flowing in the equipment ground may only be 1 mA, but the peak current can be 20 mA. A 10-mA ground fault interrupter may trip with this waveform. Note that ground fault interrupters vary in their sensitivity to peak current. A filter at the interrupter coil can solve this problem, but modifying listed equipment is not permitted. It would be incorrect to ask the user to modify the interrupter.

As power tools and consumer goods evolve, more and more line switching circuits are used. If these circuits use line filters that allow high-frequency current to flow in the equipment grounding conductor, then manufacturers must make sure that their equipment does not trip these interrupters. When users must lift the safety ground to operate their equipment, the manufacturer is liable if there is a problem. Also see Section 11.8.

8.22 CONTROLLERS

One form of load control involves turning the ac power on a percentage of each cycle. Circuits operate on positive and negative half-cycles. Triacs—gate-controlled diodes—can provide this operation. They can be used for motor speed control or light dimming, for example. After the triac is turned on, the current must return to zero before the device can be turned off. In many applications optically coupled drivers are used to control the gate. This allows the control circuit to be at any reference potential, and the triac can be associated with the power conductors.

When the triac is turned on, the voltage to the load rises in about 1 μs. The rise time depends on the nature of the power source and the load. For resistive loads the initial current demand is usually higher than for an inductive load. In both cases a short spike is apt to propagate on the power conductors back toward the service entrance.

The step function of voltage that reaches the load can radiate if the conductors are not shielded by conduit. Most motor frames are open to allow for cooling. These apertures allow for some field energy to escape. Any fields external to the conduit must return energy to the source of power via the conductors in the facility rather than inside power circuit. It is desirable to increase the rise time of the step function to the load. A motor or lamp circuit can operate well with a 100-μs delay. All that is required is a small filter at the triac output. The triac may require an input filter capacitor to ensure sufficient current flow upon turn-on.

An ideal controller has a local energy source (capacitor) and an output filter. Circuit geometry shields the region where there are rapidly changing fields. In this case, fields that could radiate are confined to a "box."

Circuits that leave the triac conducting for more than a few cycles do not create an interference problem. Gate controllers are available that operate the triacs at zero-voltage crossings. This eliminates any steep voltage leading edges. At no time is there a steep leading edge supplied to the load.

8.23   GROUND CURRENTS AND STRAY FIELDS

The multiple grounding of neutrals along a distribution system allows some power current to use the earth. At points along the system the voltage gradients are likely to be small. Where soil conditions are poor, the current will most likely follow conductors buried in the soil. This can occur in residential areas where water and gas lines abound.

Swimming pools can be a problem where the pool apron has steel not bonded to equipment ground. The water is in contact with lighting fixtures and assumes the potential of the equipment ground. The apron assumes the potential of the earth in that region. The potential difference is sensed by someone sitting on a wet apron and dangling bare feet in the water.

In a more serious vein the magnetic field from sun-spot activity can induce significant current flow in the earth. The frequencies are quite low but so are the impedances when large volumes of earth are involved. In the tundra along the Alaskan pipeline, the steel pipe is supported by pillars buried deep in the earth below the freeze line. The loops formed have the greatest loop area when the earth is frozen. Pipeline currents in excess of 400 A have been reported.

The power grid in the eastern United States begins to misbehave during sun-spot activity. When earth currents flow in ground fault interrupters, these systems detect a fault. If faults are detected in a random pattern, the power grid may be lost. If any of this low-frequency current flows in transformer coils, the transformers can saturate and the resulting magnetizing current can trip breakers. If this occurs in a random pattern, the system could be lost.

Localized ground current flows under clouds that carry charge. This current flows in building steel, buried conduit, and multiply grounded signal conductors. This current flow is usually small compared with the sun-spot activity.

An area of concern involves dairy milking machines. If any current flows through the cow from its udder to its feet, the cow is quite uncomfortable, and lawsuits have been filed against utilities. An analysis shows that the main culprit is often local equipment grounding current flow. The equipment must be kept very clean and designed so that stray current does not flow in the milking path. Placing a metal plate under the cow and connecting it to equipment ground may not resolve the issue. The cow may refuse to step on the plate if it differs in potential from the surrounding area.

A great deal of concern has been expressed over the impact of magnetic fields on human health. Tests have shown that some biological growth is influenced by small ac magnetic fields. (The field intensity of interest ranges from a few gauss to 0.01 G.) At this writing no conclusions have been reached. The magnetic field in a home result from a variety of sources. Common sources are appliances and electric blankets. Fields can result from neutral currents flowing in all of the metal conduit entering the house. Another source of magnetic field is the current flowing in overhead utility lines.

Equipment ground current in medical electronics is limited by code (UL 478). Very small currents flowing subcutaneously can be dangerous to a patient in intensive care. The current limit is 20 μA. This low limit precludes the use of conventional filter capacitors to equipment ground. Equipment ground current in portable equipment is limited to 0.5 mA (office equipment, 0.25 mA). If the equipment ground connection is violated, then any filter capacitors form a voltage divider to the equipment frame. The user then forms a circuit between this frame and earth. Limiting the equipment ground current reduces the size of the filter capacitors. This limits the current flow through the user when the equipment is not grounded.

The equipment ground current in commercial grounded equipment is not restricted by code. Military codes, however, limit equipment ground current even in grounded equipment. Lifting equipment ground connections to reduce noise is illegal and dangerous. If many pieces of equipment are involved, a serious shock hazard can exist because of the many parallel filter capacitors. A current of only 10 mA causes involuntary muscle contractions.

CHAPTER 9

CABLES AND SHIELDS

9.1 INTRODUCTION

A mathematical theory for cable performance does not exist. The task of selecting the correct cable type is not simple. Signal transmission is only part of the subject. Of importance is the coupling of external fields to the signal path. In the following discussions, the length of the cable run is very important. If the run is short, the selection of cable type may not be critical. In many applications the cable type is far less important than the nature of the shield termination.

The subject of the characteristics of practical shielding materials is also not simple. The basic ideas of reflection, absorption, and skin effect were given in Section 4.9. One example of the difficulty is material permeability. Permeability is not a constant but a function of frequency and flux level. When the skin depth equation was developed, these factors were not considered.

The transport and containment of high-frequency fields using coaxial cable requires the cable shield to be a signal conductor. The shield must be bonded at the signal source and at the signal termination. At audio frequencies the outer sheath can be used as an electrostatic shield. In this application the shield should only be connected once to the zero-reference conductor. If this conductor is grounded, then the shield is also grounded there.

9.2 BRAIDED CABLE

A common material for shielding cable is braid. It consists of fine strands of tinned copper woven into a sheath that surrounds a group of conductors. The tightness of the weave affects the electrical quality of the cable. This

162

construction allows the cable to be flexible, which is desirable in most applications.

Currents in the braid tend to follow the individual strands of wire. At audio frequencies this is of little concern because skin effect is not an issue. In other words, at low frequencies, currents tend to flow in the entire braid. This means that any E field gradient appears on the inside and outside surfaces of the shield.

At frequencies where skin effect normally limits field penetration, the braid invites currents to cross over from the outer to inner surfaces. If there is current flow on the inside surface, there is field present. This field is free to propagate in both directions along the inside of the cable. This is one way external fields can couple into the signal path.

Another view of how high-frequency fields penetrate a braided shield involves the braid apertures. The apertures are not totally independent, and thus field penetration is a function of cable length and braid density. At audio frequencies, a shield is often used to protect against electric fields. The penetration of electric fields can be described in terms of mutual capacitances. Current flowing in this capacitance enters the signal conductors and flows in the source and terminating impedances. Mutual capacitance is somewhat related to optical coverage. The looser the braid, the higher is the mutual capacitance between internal and external conductors.

Braids made of copper cannot shield directly against magnetic fields because copper has a permeability of 1. Any loop areas formed by conductors can couple to a changing magnetic field by Faraday's law. Loops of interest can involve the signal conductors or the shield and its end connections. The voltage coupled to the loop area causes current to flow based on loop impedances. If this current couples to the signal in some way, the circuit geometry must be changed to limit the magnetic coupling.

A common shielded cable is a twisted conductor pair with a braided shield. Twisting reduces differential coupling of the signal pair to external magnetic fields. The conductors might be assigned to carry signal and signal common or to carry a balanced differential signal. As the twist goes through each 360°, any coupling is nulled out, assuming that the cable is not routed near an intense field. It is impossible to provide any twisting when coaxial cable is used. This makes coaxial cable sensitive to magnetic coupling.

9.3 MULTIPLE SHIELDS

Cable is available with two outer braided shields. These shields can be in direct contact with each other or separated by a layer of insulation. The added shield improves the overall shielding performance, but often the improvement is not sufficient to justify the added cost. In one application the outer shield can be used to carry high-frequency currents and can be multiply grounded.

If the inner shield is a guard shield, it should be grounded once at the signal source. In this application the shields are insulated from each other.

When two shields are insulated, the fields that penetrate the outer shield cause current to flow in the inner shield. Coupling between shields is high because of the tight geometry. The fact that the shields are insulated from each other does not limit this coupling.

If the outer shield is used for carrying high-frequency interference, this one shield can be shared by many cables. Routing many cables in one conduit is a good solution to this problem. It is also far less expensive to use one conduit than special double-shielded cable.

9.4   FOIL-WRAPPED SHIELDS

Many standard cables use a wrap of aluminum foil to form the shield. Its shielding quality is excellent but the foil's fragile nature causes a few problems. It is not practical to expect the user to solder or weld to this foil at a connector termination. To avoid this complication, most foil-wrap shielded cables use an added drain wire. This is a bare tinned conductor that connects to the foil along the cable run. The user makes a shield connection by using this drain wire.

Aluminum foil is easily torn, so the cable must have an outer protective jacket. The foil is usually anodized on the outside surface, and the drain wire is routed inside the shield. In some cables the drain wire is routed outside the foil wrap. In this case the aluminum foil must have an external conductive surface. This is the preferred design when a drain wire is needed.

The drain conductor carries most of the shield current. If the drain conductor is inside the foil wrap, the fields associated with this current flow couple directly to the signal conductors. This means that many of the protective aspects of shielding are violated. In some applications interference coupling can disrupt or actually destroy integrated circuits in associated hardware.

Foil-wrapped cables are not often specified for applications above the audio frequency band. The foil is not a continuous conductor around the cable, and the surfaces are not cylindrical. This means that the characteristic impedance is not well defined. At a connector the drain conductor cannot provide a 360° shield termination, and this becomes an entry point for interference. In applications involving short runs this may not be of concern.

9.5   THIN-WALL SHIELDED CABLE

Cable designed to transport signals at hundreds of megahertz over any distance (cable television) requires tight control of conductor geometry. The inner conductor can be centered with a spiral of nylon cord. The shield should have a smooth uniform inner surface. If this shield is thin-wall aluminum (extruded tubing), the cable can be kept moderately flexible. The internal dielectric is mostly air, which limits distributed dielectric absorption, which in turn causes unwanted reflections to occur.

Flexible thin-wall coaxial cable is commercially available. Cable flexibility is achieved by keeping the cable diameter small and slightly corrugating the outer conductor. This cable is more expensive than braided cable but is far less susceptible to external interference. This assumes that the sheath is properly terminated.

9.6 TRANSFER IMPEDANCE

Transfer impedance is a measure of shielding effectivity. A current is caused to flow in the shield. The cable under test is terminated at both ends. A voltmeter measures the signal appearing at one of the terminations. The transfer impedance (in ohms) is the ratio of voltage to the current. The measure is normalized for a cable length of 1 m. Figure 9.1 indicates the nature of the test. One-half of the coupled voltage appears at each end of the cable. The impedance is thus given by the ratio of twice the voltage at one termination to the current flow.

The transfer impedance is given in units of dB ohms/m, where 0 dB ohms = 1 ohm and 20 dB ohms = 10 ohms. The transfer impedance for typical cables is given in Figure 9.2. Note that for most braided cables the impedance rises as the frequency increases. Multiple shields are somewhat better than single shields. A solid shield provides the best performance at high frequencies.

In calculating the transfer impedance for a typical cable, the coupling is proportional to cable length up to a half-wavelength. Above this length the coupling actually decreases. For a worst-case approach in analysis, the half-wavelength value is used when the cable length is greater than a half-wavelength.

9.7 SHIELD TERMINATIONS

A connector is an aperture that can allow interference into an enclosure. The conductors that use the connector can also transport interference into the enclosure. If the conductors and connector are shielded properly, interference cannot enter at this point. The real test of shielding occurs when the equipment is subjected to a very noisy environment. The chattering relay test is one such environment. A relay and its contacts are set up as a buzzer. The circuit conductors are wound around an interconnecting shielded cable. This buzzer circuit couples a very broad spectrum of interfering current into the shield. Another very noisy source is an electric hand drill.

When a shield is terminated with a short piece of wire, any shield current flow is concentrated in the wire. This short stub of wire can be thought of as

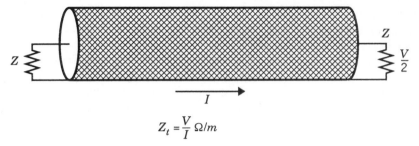

$$Z_t = \frac{V}{I} \ \Omega/m$$

Figure 9.1 The transfer impedance test.

an inductance. The field associated with this inductance is located right at the entrance to the connector. The result is that every lead entering the connector couples to this field and contaminates the system. This energy also goes backwards through the cable, contaminating the signal source and its power.

The greatest field coupling occurs when the shield is terminated through one pin on the connector. This brings interference directly into the enclosure.

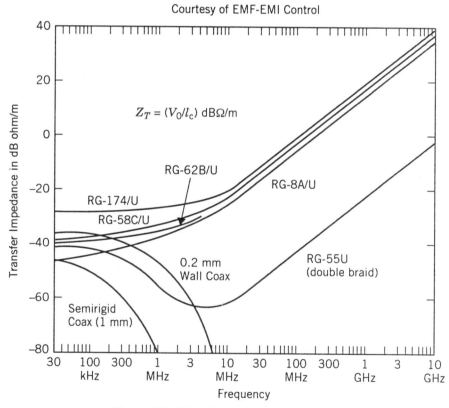

Figure 9.2 Typical transfer impedance curves.

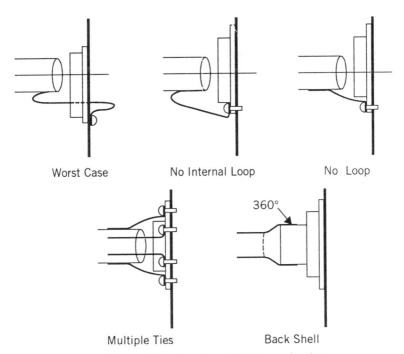

Figure 9.3 A series of shield terminations.

It is somewhat better if the shield is terminated outside the bulkhead. Unfortunately, the lead inductance will still develop fields that enter the box. If the braid itself is used as a conductor, the interference is somewhat reduced. The only solution that keeps interference out of the box is a full 360° termination of the shield. Usually this requires a back-shell connector. The shield current is then restricted to flow outside the shield and bulkhead. Fields thus cannot concentrate near an aperture or near conductors entering through a connector.

It is important to note that the bond between the shield and bulkhead can be microhms whether the shield itself is used or a back-shell connector is used. One method develops a field that enters the system, and the other does not. This is in the realm of circuit geometry, not circuit theory. It is important to know how connections are made as well as where they are made. A sequence of connections is shown in Figure 9.3. Each step represents an improvement.

9.8 TWISTED-PAIR UNSHIELDED CABLE

Shielded twisted-pair cable may not perform as well as simple twisted-pair cable. A shield actually invites interference currents to flow near the conductors. At about 100 MHz a braided shield is not very effective. If the conductor pair is used to transport a balanced signal, common-mode coupling can be eliminated by rejecting common-mode content at the receiver circuitry. Differential interference coupling is limited by twisting the conductors.

An example of twisted-pair usage involves the transport of standard color video signals over miles of cable (typical bell wire). The hardware used must provide pre- and postemphasis of signals. Repeaters are required at mile intervals. The conductor insulation used must not have significant dielectric absorption, or compensation for signal loss is not practical. The wire pair must be continuous without stubs.

Another example of open-pair usage involves the transport of digital logic. In one installation a bus consisting of 100 conductor pairs was successfully operated at a clock rate of 110 MHz over a 70-ft distance. This is no different than using a long run of unshielded ribbon cable.

9.9   RIBBON CABLES

Ribbon cables are simply a group of conductors running parallel, like the strings on a piano. The conductors are separated by their insulation and are usually color coded for easy identification. The pitch (spacing) of the conductors is controlled with several available standard pitches. A variety of connectors is available that contact the ribbon conductors by piercing the insulation. These connectors allow terminations, stubbing, and extensions. The ribbon width can vary from a few conductors to many dozen. Ribbon cables make multiple connections quite manageable. This situation is typical for digital systems.

Ribbon cables are available with a ground plane on one side, a wraparound shield, individual isolated shield, or shared shields. These cables are usually used to transport logic in a coaxial manner. A coaxial treatment of signal limits radiation and makes the signal path less susceptible to external radiation. If the shield is aluminum foil, all of the problems discussed in Section 9.3 can arise. Hardware to properly terminate a foil wrap is available, but it is expensive.

Ribbon cables are available with twisted pairs. Twisting can be used to limit differential coupling from external fields or cross-coupling between ribbon cable pairs. Not all pairs need be twisted. If adjacent pairs are cross-coupling, it is preferable to twist one of the pairs to reduce this coupling. Twisting both pairs with the same number of twists per foot may not be effective, even if the twisting is in opposite directions. An examination of loop areas and their orientations illustrates why this is true.

9.10   USE OF GROUNDING CONDUCTORS IN RIBBON CABLE

Consider a printed circuit board handling logic signals. The traces on the circuit board can be viewed as transmission lines using the ground plane as a return conductor. The fields associated with this transport are concentrated under the trace. This geometry greatly limits crosstalk and susceptibility from external fields.

This same philosophy should be used when logic signals leave a circuit board via a ribbon cable. Each signal must be transported over a known transmission line that is continuous through the connector and onto the ribbon cable. This implies that a continuous ground path should be provided next to each signal path. If this ground path is not provided, the resulting loop areas can radiate and allow crosstalk. These same loop areas increase the system's susceptibility to external interference.

A single shared ground conductor through a connector is not recommended. If ribbon cable is used to carry logic, many conductors should be assigned as grounds. This parallels the printed circuit board concept where a ground (return conductor) is available under each trace. On a ribbon cable the minimum approach is to place a ground conductor next to each signal conductor. A typical configuration might be *sgssgssgs*, where *g* is ground and *s* is signal. All grounds should be bonded separately to the ground planes at both ends of the ribbon cable run. In effect, these grounded conductors form a bridge between the two ground planes. Multiple ground connections should not be viewed as ground loops. After all, a ground plane is nothing more than an infinite number of ground loops. The field for each signal is located between the signal conductor and the nearest ground. This field geometry requires the least field energy to transport the signal.

If the ribbon cable has an integral ground plane, it should be bonded to form a continuous ground plane between the signal sources and their termination. If this is not done, the ribbon ground plane loses its effectivity. If a ground plane is provided, one must assume that it is needed. Terminating ground planes and adding pins to terminate ground conductors increase size. Experience indicates whether this extra cost is warranted.

It is poor practice to coil up excess coaxial cable because it is susceptible to coupling from interfering fields. The best way to route ribbon cable is on a conductive surface, preferably a ground plane. If hardware is provided to hold the cable on the ground plane, the cable will stay there. Cables that crisscross an area in an undefinable manner can be susceptible to interference coupling. Careful cable routing does not cost much, and it makes a product look neat. This practice reduces any chance of common-mode coupling.

9.11 PLATED SHIELDS

Thin sheets of conducting material can be deposited on nonconducting surfaces. Techniques include sputtering, electroless plating, spraying, and brushing. Sputtering is done in a vacuum where the material to be deposited is vaporized. Sputtering is basically a line-of-sight operation. The item to be plated is often rotated under the sputtering source. Even with rotation it may be difficult to reach certain internal points. Sputtering has the advantage of layering, where each layer can be a different material. It may be difficult to obtain a uniform plating thickness. In most applications the thickness need only be adequate.

Electroless plating requires the item to be dipped in a series of chemical baths. The process is sensitive to the type of plastic involved. For example, not all nylon can be plated. This form of plating deposits material uniformly on all surfaces even if they are hidden.

Plated copper oxidizes very quickly. A plated surface must be quickly treated, or a second metal must be deposited. If this is not done, the surface soon becomes unusable. After a layer of copper has been deposited, the second layer can be added by conventional electroplating.

9.12   PAINTED SHIELDS

A variety of conductive paints can be brushed or sprayed on nonconductive surfaces. In most cases the metal content is not held in suspension, which requires that the paint be agitated during application. The surfaces must be thoroughly dry before any accurate measure of conductivity can be made. When the surface is not dry, the metallic particles are not in intimate contact with each other.

9.13   ELECTRICAL CONTACTS TO THIN SHIELDS

Making contact with a thin conductive surface can pose a problem. Contacts over a large surface area are suggested. A strip of copper can be bonded to the surface with a conductive glue. Screws and washers tend to scrape the surface, which can result in an unreliable connection. Conductive gasket material can be used to provide a connection to a thin shield. Gaskets can be used to connect a plastic cover to a housing, making the conductive surface continuous. If these contacts are discontinuous, the resulting apertures can make the shield ineffective.

A conductive surface need not be grounded to be an effective reflector of field energy. Connections to circuit common may be required to avoid coupling noise electrostatically to the circuit. Safety grounding is necessary if the surface might contact any power.

Conductive strands or particles can be added to a plastic mix before molding. If these strands are sufficiently dense, they will contact with each other and form a shield. There is usually a critical density below which this shielding will not work. Metal material in the mix can limit the life of the mold, making this process expensive. Making contact with the shielding material is difficult because it is buried in the plastic. Hardware should be molded into the plastic part so that a guaranteed connection is brought out. If a cabinet comprises several mating parts, this approach leaves apertures. To avoid apertures, all mating surfaces must somehow be made conductive and bonded together.

9.14 MAGNETIC SHIELDS

Materials with permeability can be used as magnetic shields. However, at frequencies involving power and its harmonics, shielding is a near-field problem. The wave impedance for a near-inductive field is quite low. This means that the field will not be easily reflected. Reflection is analogous to a transmission line mismatch. A low-impedance wave impinging a low-impedance surface is only partially reflected. The energy entering the metal is attenuated mainly by losses. Several inches of steel may be required to provide 40 dB of attenuation at 60 Hz. Some of the attenuation is caused by reflection, but most of it is simply skin effect.

It is impractical in most cases to provide massive steel structures to attenuate magnetic fields. Since the skin depth process is a function of permeability, it is desirable to use a very-high-permeability material as a shield. Magnetic alloys are available with very high permeability at power frequencies. One of the best-known materials is μ-metal. This alloy is annealed in a hydrogen atmosphere in the presence of a large magnetic field. The high permeability is lost, however, if the material is mechanically disturbed by bending, drilling, or heating, for example. This means the part must be formed first and then annealed to achieve maximum permeability. The annealing process limits the size of the final item and is also expensive.

Unfortunately, μ-metal saturates very easily—that is, it is ineffective near large magnetic fields. It is fine to use it around a cathode ray tube to redirect stray magnetic fields or the earth's field. It would be inappropriate to use it near a power transformer.

When a power transformer must be shielded, the winding assembly can be mounted in a nested set of cans and lids. The inner can is often ordinary steel, which will not saturate. The second can is often copper, and the third can might be a high-permeability steel. The lids are necessary to help contain the field. A small hole is used to bring out leads. It is easy to see that overheating could be a serious problem as heat is sealed in. Designs of this type rarely handle a load of more than a few watts. It is easier to design a transformer to operate at low flux density to reduce external fields than to pay for this exotic shielding.

Rather than shield an enclosed area against a magnetic field, it may be prudent to reshape the field in the area of coupling. A piece of magnetic material will alter the character of a magnetic field. The field will reconfigure to store a smaller amount of energy. If the field density is reduced in the area of coupling, the interference is also reduced.

A steel band around a connector can limit the magnetic field near the connector pins. This can solve the problem of coupling where the pin spacing provides the coupling area. This reshaping of the magnetic field is shown in Figure 9.4.

9.15 MAGNETIC MATERIALS AT HIGH FREQUENCIES

The magnetic core materials used in power transformers are usually laminated. Laminations break up the eddy current paths, thus limiting the heating of the core. The power to heat the core adds to the magnetizing current, and a high magnetizing current is undesirable in a power transformer: 60-Hz transformers often use 14-mil-thick laminations, and 400-Hz transformers use 6-mil-thick laminations. It is not practical to stack laminations if they are thinner than 6 mils.

Audio frequency transformers handle a wide range of frequencies. If the lowest frequency is 20 Hz, this defines the maximum flux level. At 2 kHz the

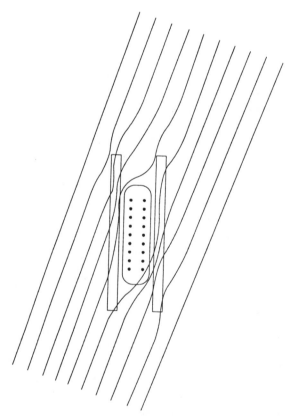

Figure 9.4 Using steel to reshape a magnetic field.

flux level is only 1% of the 20-Hz case. At this low flux level the permeability may not be very high, but, on the other hand, eddy current losses are very low. This means that very thin laminations are not required to operate a transformer over the audio frequency band.

Powdered iron cores are available for applications in high-frequency transformers. Different magnetic alloys are imbedded in a ceramic filler. The ceramic filler insulates the magnetic domains from each other, which limits the eddy current losses. Cores are available in a variety of sizes and permeabilities to accommodate many user requirements. The core is an insulator and does not reflect a low-frequency E field. These cores can provide moderate permeability at frequencies up to and above 1 MHz. This core material is known as ferrite.

A variety of binding materials is available to separate the magnetic domains of a magnetic material. A tape can be formed with the magnetic materials bonded to one surface. The tape is flexible and can be wrapped around a cable, for example. This wrap forms a magnetic shield that can deflect an external magnetic field. The field follows the high-permeability path rather than cutting through the cable. It is possible to obtain 20 to 30 dB of shielding effectivity from this type of wrap.

9.16 HIGH-FREQUENCY INDUCTORS

Magnetic materials are available in many shapes and sizes. Small cylinders (beads) are used as inductors. Toroids can be used for inductors or transformer cores. Large blocks of materials can be clamped over a cable to form a one-turn inductor.

Ferrite materials are hard and brittle. Mounting hardware is available and should be used where practical. Cores should be mounted so that they cannot be stressed during manufacture, shipment, or usage.

Inductors used in passive filters must have low temperature coefficients, be stable over time, and have reproducible characteristics. Cores for this application often have a distributed gap. Ferrite pot cores are available with an adjustable gap so that the inductance can be mechanically set. The gap adjustment range is limited. Cores with different gap ranges are available.

It is often desirable to limit high-frequency current flow by placing a circuit conductor through a small ferrite bead. If the inner diameter permits, the conductor can form a coil of several turns, which raises the inductance. Inductance is proportional to the square of the number of turns. Because of the small amount of magnetic material involved, the resulting inductance is usually quite small. The self-capacitance of the turns limits the highest application frequency. Both 10 nH and 2 pF resonate near 1 GHz. Thus, it is often better to add cores than to add turns.

Beads placed on conductors can be easily saturated if the conductor carries low frequency or dc current. The maximum H field in a core might be 0.5 At/m.

If the core has a circumference of 2 cm, the maximum one-turn current is 10 mA. For two turns the maximum current is 5 mA. A core with an effective gap can tolerate more ampere turns before saturating. The saturation characteristics of a core are usually specified by the manufacturer.

At frequencies above a few megahertz the magnetic domain losses begin to dominate. As the frequency increases, the bead begins to look like a lossy element rather than an inductor. This simply means that the impedance presented by the component does not increase linearly with frequency. If the impedance is high enough to block interference, then the nature of the impedance is not important. This lossy character is often represented in a circuit by a shunt resistor. The impedance cannot rise to a value greater than this shunt value of resistance.

The impedance presented by a single-turn bead may not rise above 20 ohms. A series of beads can be used to further raise the inductance. If the inductance of one bead is 30 nH, this 20 Ω reactance occurs at 1 GHz. This means that one bead is ineffective in circuits with impedances greater than 20 ohms. The bead is also ineffective unless the interference is in this frequency range. Beads that reduce crosstalk might be simply spacing the leads. This benefit could be provided just as well with plastic beads.

Common-mode current flow in a group of conductors can be reduced by threading the conductors through a toroidal core. In most circuits the average current is zero, and saturation is not a consideration. If the cable is large, blocks of ferrite material can be clamped around it. These blocks need to be protected to avoid damage during handling. A toroid used in this way is often called a balun—technically a transformer to convert a *bal*anced signal to an *un*balanced signal. Baluns are used in power transmission where the cable is used to transport a balanced signal and the antenna load is unbalanced.

Balanced power transmission has the advantage that the sheath does not carry the load current. This technique limits the power leakage out of the cable. At high power levels this can be an important consideration. The sheath must still be terminated at each end of the cable.

9.17   SHIELDING A VISUAL DISPLAY TERMINAL

A cathode ray tube (CRT) represents a large aperture, which allows radiation to leave an enclosure and allows field energy to enter. Radiation can come from electronics in the enclosure or from the electron beam itself. When the beam is deflected in a step manner, radiation occurs. When the beam strikes the face of the CRT, a radiating field is generated by the partial loop of current formed. This field is near field and difficult to shield.

The field pattern generated is directly related to the writing on the screen. With the correct equipment and software it is possible for an eavesdropper to reconstruct all of the data displayed on a CRT. In areas where security is important this fact is dismaying.

A wire grid can be used to shield the CRT aperture. Unfortunately, a grid structure is associated with the character generation scheme used on most display terminals. This combination of grid patterns forms moire patterns that can be quite disturbing to the viewer. It is not practical to limit this beat phenomena by selecting grid pattern dimensions.

A popular solution is to screen the CRT with a sheet of glass that has a conductive surface but no grid structure. One way to obtain a conductive surface is to quick-cool the glass while a layer of gold is suspended in the melt. These glass screens come with a conductive ring so that there is no peripheral aperture when they are mounted.

A relationship does exist between opacity and shielding effectivity. A good shield ends up attenuating light as well as the unwanted electromagnetic field. This solution is thus a compromise. Other solutions exist. The entire viewing area can be screened, or the CRT can be viewed through an optical system. If the optics forms a waveguide, the offending field can be well attenuated.

New technologies using flat displays can greatly reduce this radiation problem. When the screen is not refreshed on a continuous basis and a high-voltage electron beam is not used, the problem simply vanishes.

The low-frequency fields generated by the deflection coils in monitor designs are near-induction fields. The biological impact of exposure to these fields over extended periods is unknown. The fear that there might be problems from long-term exposure has resulted in specifications that limit the intensity of these magnetic fields. These limitations have resulted in deflection systems that greatly reduce these external fields. Control is accomplished by providing a low-reluctance return path for the flux at the deflection coils. This approach is used in monitor designs to meet the new European standards for magnetic leakage flux levels near the monitor.

The Swedish standards, which seem to prevail, cover the following frequency bands. Band 1 covers the range 5 Hz to 2 kHz, and band 2 covers 2 kHz to 400 kHz. The latest specifications are defined by the Swedish Conference of Federal Employees. In band 1 the field strength is limited to 200 nT (2 mG) at a distance of 50 cm around the monitor and 30 cm in front of the monitor. In band 2 the field strength is limited to 25 nT (0.25 mG) for a distance of 50 cm around the monitor.

In older monitor designs the external flux can be attenuated by adding a magnetic path around the deflection coils. This could be a ferrite core or magnetic tape. The flux can also be attenuated outside of the monitor case by adding magnetic material around the monitor. This approach is less effective because the face of the monitor cannot be covered. To help limit the flux crossing this area, one can place a shorted turn around the monitor opening.

The European standards also limit the E field strength 50 cm in front of the monitor. In band 1 the field is limited to 25 V/m, in band 2 to 2.5 V/m. A conductive glass on the front of the CRT provides adequate E field shielding. This conduction also limits any charge buildup on the outside surface. The standards limit the surface potential to 500 V at any point on the glass.

9.18   CABLE-TO-CABLE COUPLING

The currents or voltage in one cable pair can couple signals to a second cable pair. The coupling can be differential or common mode in nature. If the field couples to the area between the signal conductors, the interference is differential. If the field couples between a signal cable and some reference conductor (earth), the interference is common mode.

The general solution to this type of problem is complex because the two cables are distributed systems. To simplify the analysis, it is again necessary to take a worst-case approach. For example, the field is assumed to be oriented for maximum coupling. The coupling is assumed to be proportional to cable length, even though some form of cancellation may be present. It is convenient to solve the problem in two parts. First, assume the coupling is capacitive and solve the problem by using E fields. Second, assume the coupling is magnetic and solve the problem by using H fields. If one of the solutions dominates, discard the other calculation.

The E field solution involves the capacitance between two parallel conductors. For small-diameter wires the capacitance is

$$\frac{C = 3.677}{\log(2D/d)} \text{ pF/ft} \tag{9.1}$$

The voltage on one conductor causes current to flow in the second conductor. The coupled voltage is calculated based on the impedance of the circuit at the frequency of interest. If terminating impedances are involved, use the parallel combination of impedance. The source impedance is of no concern.

The H field solution involves loop area. The interference voltage is solved using Equation (2.15). The H field is proportional to the interfering current. The lower the impedance in the "culprit" circuit, the greater the interference will be in the "victim" circuit.

If the two calculations of interference are about equal, the worst-case analysis requires the assumption that the interfering processes are simply additive.

In most cases the interfering process is a pulse or nonsinusoidal signal. The sine wave signal to use in calculation has a frequency of $1/\pi\tau_r$ and an amplitude equal to the pulse height or signal amplitude of the interfering signal. The result is an amplitude, not a waveform. In most cases the magnitude of the interference is all that is needed.

CHAPTER 10

PRINTED CIRCUITS

10.1 INTRODUCTION

Significant advances are constantly being made in circuit board technology. Over the years trace width has dropped, and the number of board layers has increased. Equipment is available to automatically position and solder surface-mounted components with great accuracy. Chip carriers have proliferated, making it possible to mount very large and complex integrated circuits. All of these advances are the direct result of the digital explosion that is moving the electronics world.

One driving force in board design is the speed of digital logic. Circuits with high clock rates cannot be operated without a ground plane. A single board could have over 1000 interconnections. To handle them, multilevel boards have become commonplace. Another aspect of printed circuit board design is limiting radiation to meet various standards. Radiation is controlled by the use of ground planes, although this is not the entire story.

Circuit designers often do not understand how a ground plane functions. Using standard design approaches keeps the engineer out of trouble. When the circuits leave the board domain, errors are made that can allow radiation or increase susceptibility to external interference. These topics are discussed in this chapter.

Clock rates have steadily increased from a few megahertz to above 100 MHz. Gigahertz clock speeds will be next. As the speed increases, the technology must change to make these speeds possible. The questions that must be asked relate to how the technology must change. Are the present approaches to printed circuit board design adequate? Can traces take right-angle turns?

Are decoupling capacitors adequate? Let's use highway engineering as an analogy. What would it take to change a highway so that cars could drive safely at 120 mph instead of 60 mph? This is only a factor of 2 change in speed. Would one dare ask for an order-of-magnitude change? Yet the digital world thinks nothing of increasing speeds by factors of 10.

It is obvious for highways that changes must be made. For printed circuit boards the approaches have remained nearly the same, yet speeds have increased by orders of magnitude. There is trouble ahead. The changes that must be made relate to the physics of field energy transport. This is the only viewpoint that can lead the way.

10.2   CIRCUIT TRACES

It is important to start this discussion with an understanding that all signals and power are transported by fields. The conductors simply direct where the signal and power can flow. The circuit traces are transmission lines with the return conductor as the ground plane. Most of the field energy is contained in the space between the trace and the ground plane.

Logic circuits are solid-state switches that connect or disconnect the power supply to traces for transmission to another logic element. The switch impedance and the input impedance of the logic element are nonlinear, which means that the term *impedance* is technically not correct. A frequent assumption is that the source impedance is near zero and the terminating impedance is some average resistance. This impedance is a function of the logic family involved. The presence of a power voltage on a logic input is referred to as a logic 1. The absence of a power voltage (the presence of a ground reference potential) is called a logic 0.

The switch that connects the input of a transmission line to the power voltage or common is not perfect. On a 5-V supply the logic signals might transition from 0.2 to 4.8 V. The power supply potentials are often referred to as "rails." The logic signal is said to be 0.2 V away from the rail.

Synchronous logic requires a clock signal during logical activity. Each logical element responds to the signals on its inputs when the clock signal makes a transition. The response may be on a leading edge (positive transition of the clock) or on a falling edge (negative transition of the clock). The logical element takes time to respond to the new states. In a short time new logic signals appear on its outputs. These signals are then transported by circuit traces to other circuit elements. The transported signals may reflect several times on the transmission lines since the lines are not accurately terminated. The hope is that within one clock time (period) all transient activity will have sufficiently attenuated so that a logical error will not be made.

Many logical devices will not function correctly unless the leading edge of the clock signal rises or falls in a short-enough time. If the leading edge is too slow, parts of the internal logic can transition early, which causes a malfunction.

10.3 DECOUPLING CAPACITORS

Energy sent forward on every transmission line must come from somewhere. If this energy is not supplied locally, the nearest source of energy is the power supply. The transmission line back to the power supply might have a characteristic impedance of 50 ohms. If 10 logic switches request energy at the same time and the logic transmission line impedances are each 50 ohms, the voltage drops to about 10% of the initial value. If the power supply drops to this value, then all logic is affected and will malfunction.

Acceptable delays in providing energy for logic transmission depend on the clock rate. A 33-MHz clock rate requires a logic cycle every 30 ns. If the clock leading edge rises in 5 ns, if logic delays are 10 ns, and transmission line settling times are 5 ns, the power supply delay is limited to 10 ns. Time is needed for the energy request to get from the logic element to the energy source, for the energy source to release some energy, and for this energy to get back to the logic element. If these delays exceed 10 ns, the logic will not function. Time is also needed for clock transport and signal transport. It is good design practice to design with a reasonable timing margin so that logic will function even if the delays change. It is easy to see that in high-speed logic, every delay is important.

It is standard practice to provide a source of energy near each logic circuit. The time delays involved in supplying energy to a logic transmission line are limited to the nature of the storing device and to the conductor geometry. Decoupling capacitors are intended to perform this function. In circuits with clock rates of 3 MHz, one decoupling capacitor may be required for every three or four ICs. Capacitors are available that mount directly under an IC.

The lead geometry inside the capacitor, the leads to the capacitor, and the leads on the integrated circuit define a loop. This loop implies inductance, which is in series with the capacitor: the larger the loop, the larger the inductance. It is usual to describe this as a series resonant tank circuit. At low frequencies the circuit looks like a capacitor. Above the natural frequency the circuit looks like an inductance: the larger the capacitor, the lower the natural frequency.

The natural frequency of a 1000-pF capacitor and an inductance of 1 nH is 159 MHz. A decoupling capacitor of 0.01 μF has a natural frequency of 15.9 MHz. Hence, a larger capacitor is not necessarily better. In the decoupling process a larger number of small-valued capacitors is preferred. Capacitors should be mounted with short leads yet long enough to provide needed strain relief.

Capacitor dielectrics differ, and they release their stored energy in different lengths of time. In this sense, the capacitor is a nonlinear element. In selecting capacitors for decoupling applications the character of the dielectric should be specified. Capacitor manufacturers provide information on the performance of various capacitor types intended for decoupling applications. Electrolytic and tantalum capacitors can be used as general energy storage for an entire board. It is not recommended that they be used for local decoupling.

Surface-mounted capacitors form the smallest loop areas and thus have the lowest series inductance. In high-clock-rate designs this component style is recommended for decoupling application.

A ground and power plane represent a local energy source. These parallel plates can be viewed as a capacitor or as a transmission line. The characteristic impedance as viewed from one edge of the board might be below 1 ohm. The source impedance as viewed from an integrated circuit is higher. This local source of energy is in parallel with any decoupling capacitors. In the first fraction of a nanosecond this may be the only significant energy available.

Wire-wrap techniques can be used when clock rates are not too demanding. Short production runs of small boards (6 × 8 in.) can be built this way. If this approach tends to malfunction, it may be necessary to add a ground grid connecting all IC commons together. This approximates a ground plane not always found with wire-wrap designs. If a ground plane is available, it should be used. Long trace runs should be connected first so that they can be routed on the ground plane. Neat right-angle trace runs add loop area and should be avoided. Decoupling capacitors are still necessary. If wire-wrap programming is available, algorithms to minimize trace length should be used. To help the program find a viable solution, complex circuits should be handled in blocks. Integrated circuit locations should be decided manually so that there is a general flow to the circuits.

10.4   MULTILAYER BOARDS

Applications with numerous interconnections need many board layers to keep the component density high. This consideration is important because board size determines product size and weight and the eventual price of the unit.

A ground plane serves as a return conductor for every trace. This construction technique limits radiation from the board, which is proportional to signal loop area. The loops that can radiate occur mainly at the IC header, where the legs of the IC are involved, and at the decoupling capacitors. Loops formed at connectors are discussed later.

The supply voltage to a group of ICs need not be provided from a power plane. Power and signal traces can be run together in any mix. The single ground plane together with properly distributed decoupling capacitors provides an adequate design. If the supply voltage traces must be too small, then *IR* drops in the supply traces may be unacceptable. The added power plane avoids these voltage drops. Decoupling capacitors are still required when a power plane is provided. An added power plane is electrically another ground plane.

Traces that stay on opposite sides of a ground or power plane do not crosstalk. The fields for signal transport are located on opposite sides of the ground plane. Skin effect keeps the fields apart even inside the ground plane. The fields associated with traces that use the space between ground planes (or ground and power planes) are definitely contained and do not radiate. An outside trace does have some field at a distance from the board and can radiate.

Some radiation can take place from the edges of a printed circuit board. This radiation can be reduced by keeping traces away from the edges and running the ground planes out to the edges. It is preferable to not use a power plane for the top and bottom layers. Even if the plane is conformally coated, there is some danger of shorting it to ground if a tool is dropped on the board.

It is impractical to breadboard high-speed logic designs involving multilayer boards. The only practical approach is to build the final circuit. To avoid costly mistakes the design should be simulated on a computer. All the logic can be exercised, including all expected time delays on and off the board. If testing is successful, the design can be built. This type of computer simulation, however, does not indicate problems in radiation or susceptibility.

10.5   MIXING ANALOG AND DIGITAL CIRCUITRY

Should separate logic and analog grounds be used? Designers often think that the two grounds should connect once at the digital-to-analog interface. A glance at any circuit approach requires signals to use this single connection. At high frequencies such a narrow path represents an inductance. There may be little voltage gradient in the ground planes, but at this most critical point there is a voltage drop.

The consensus among many designers is that one ground plane should be used. Analog and digital processes can be separated when the signal field locations are known. These fields involve power and signal. When the fields occupy different regions of space, the two types of signals will not crosstalk. If the geometries are carefully controlled, even radiated field energy is not an issue. A single-ground-plane approach is thus highly dependent on circuit layout. Every lead that defines the path for signal fields must be considered.

The use of one connector for analog and digital signals makes the problem difficult. If the logic and analog circuits share the same field space, a solution may not exist. Separation of fields must be considered early in product design.

The mix of analog and digital circuitry often involves an A/D (analog-to-digital) converter. A 16-bit converter with a 10-V input signal must resolve voltages as small as 150 μV. This level of signal can result from logic current flow in a ground plane or logic signals coupling into an analog circuit. Even on a small ground plane this level of error can occur. A higher-resolution A/D converter requires an even smaller error limit.

It is good practice to use an A/D converter with an integral differential input. The input leads sense an analog difference and ignore any common-mode (ground potential difference) signal. This differential amplifier is also called a forward referencing amplifier. The nature of the source impedance is the key to common-mode rejection. This performance should be part of the manufacturer's specification. If necessary, this type of analog circuitry can be added. See Section 5.19.

10.6 CONNECTORS AND DIGITAL CIRCUITS

Various connectors are available for use on printed circuit boards. Boards that plug into main boards (motherboards) or back planes often plug into edge connectors. Ribbon cable connectors and various coaxial cable terminations are also available.

The extension of ground planes onto the ribbon cable requires terminating many ground conductors. These connections should be made directly to the ground plane if available. The shell of a coaxial connector should make a 360° bond to the ground plane if possible. If soldering is not acceptable, then a conducting gasket should be used.

Gold is the preferred contact material for low-level analog signals. The amount of gold depends on the number of insertions expected over the life of the product. Oxides forming on the contacts are not a problem if there is sufficient voltage across the contact. Circuits rarely exercised or with very small signals often malfunction when the contacts are not gold.

10.7 ANALOG CIRCUIT LAYOUTS

Analog circuits can be built without ground planes. Experience indicates that a bandwidth of 100 kHz, 2 μV rms rti noise, and gains to 10,000 can be built without shielded conductors or ground planes. This includes on-board power transformers. If on-board microprocessors are used, these circuits can be operated at clock rates around 1 MHz and interference can be controlled. It is practical to turn the clock off when processing is not required.

The analog layout approach is simple. All signal paths and signal return paths are given priority in layout. Paths are kept short and form minimum loop areas. Summing points are kept very tight with the shortest lead length. The input circuitry has the highest priority. It is kept away from transformers, from clock lines and from fast-moving air. Input leads are guarded along their entire run. Voltages are routed from the output stage forward. This keeps output current from flowing in input common leads.

10.8 SURFACE EFFECTS IN HIGH-FREQUENCY CIRCUITS

Prototype circuits often behave differently from a manufactured item, particularly for circuits operating above 10 MHz. At first glance the circuit layouts appear to be the same, but the differences can be subtle. A connector on a prototype may be soldered to a ground plane but riveted in production. The riveting does not provide a quality bond, and the connection is inadequate. Traces are often very irregular. Skin effect forces the current to flow on the surface. If the current path is "bumpy," circuit inductance increases and performance changes. These are perfect examples of where the circuit is an exercise in geometry. To the untrained eye these subtleties are just not visible.

10.9 LIMITING EQUIPMENT GROUND CURRENT AT HIGH FREQUENCIES

Switching power supplies use transistors as switches. The collectors on transistors or the drains on FET devices must dissipate a great deal of heat. Even if the "on" resistance is 0.2 ohm, a 10-A current requires the semiconductor to dissipate 10 W if turned "on" half the time. It is often necessary to conduct this heat onto a large metal chassis than to use a heat sink that is coupled to air.

An insulator is required between the semiconductor and its mounting surface. Rarely can the semiconductor mounting be used to ground the circuit. This insulator should conduct heat but not current. Various heat-sinking compounds can be spread on the insulator to improve heat conduction. However, the equipment ground, the insulator, and the semiconductor case form a capacitor. If the collector of a transistor changes from 0 to 150 V in 0.1 μs and the capacitance is 30 pF, the current flow is $C(dV/dt)$, or 4.5 mA. Spikes caused by switching are now free to roam the facility.

Mounting insulators are available with a center conductor to act as a guard shield and cause most of the reactive current to return to the circuit rather than flow in the equipment ground. The mutual capacitance to the equipment ground can be reduced to perhaps 2 pF. The equipment ground current is now only 30 μA. This circuit is shown in Figure 10.1.

10.10 RADIATION FROM PRINTED CIRCUIT BOARDS

The loop areas that radiate from a printed circuit board involve the legs on the IC package and the conductors to the decoupling capacitors. These areas are small, but the number of loops and logic switch closures per clock cycle can multiply the effect. The circuit traces radiate, but the effect is usually second-order.

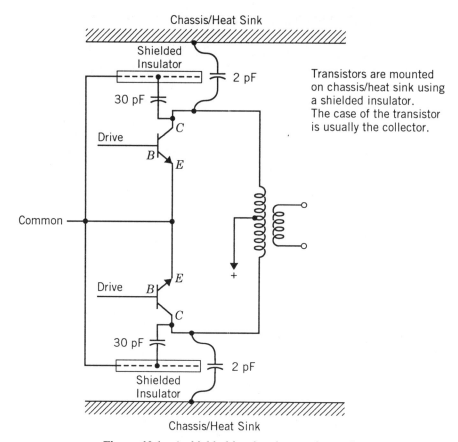

Figure 10.1 A shielded insulated mounting pad.

A worst-case analysis is the only one that makes sense. The radiation from each source is assumed to be additive. All sources are assumed to be oriented to add to the total radiation. An exact analysis is not practical.

The radiation from a single square loop of wire carrying a sinusoidal current is given in Figure 10.2. Data are given for a square 1 cm on a side (an area of 1 cm²) for various terminating impedances when driven at 1 V rms. The radiation level is given (in μV/m) for distances of 1 and 10 m. At higher frequencies the radiation from a 30-ohm termination at 1 m is the same as the radiation from a 10-ohm circuit at 3 m. For radiation levels at greater distances the field strength falls off proportionally to distance.

To use Figure 10.2, assume that radiation is proportional to area, voltage, number of loops, and logic transitions per clock time for each integrated circuit. The frequency that must be used is $1/\pi\tau_r$, where τ_r is the rise time.

Assume a board with 30 ICs. The logic family is TTL operating at 5 V. The loop areas to the logic and from the capacitors are 0.1 cm. On the average,

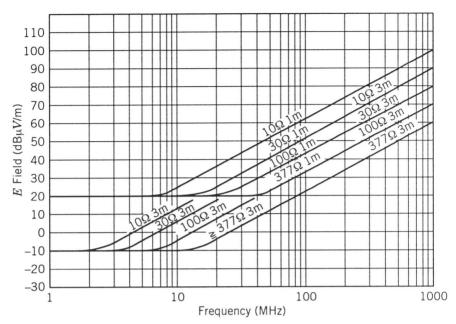

Figure 10.2 Radiation from loops of wire.

10 logic transitions occur per clock time per IC. Assume the impedance level is 100 ohms. Enter the curves at 100 ohms, 3 m at a frequency of 30 MHz. The normalized field strength is 12 dBμV/m. This figure needs to be multiplied by 5 (voltage), by 30 (number of ICs), by 10 (number of transitions), and by 0.1 (loop area). The radiation in this case is 150 × 12 dBμV/m = 55 dBμV/m.

Radiation is proportional to loop area. The largest areas likely occur at connectors and on interconnecting cable. The loop areas not only radiate but they can couple to board radiation and transport this energy superposed on signal.

10.11 RADIATION SPECTRUM AND PRINTED CIRCUIT BOARDS

Radiation from a product is often analyzed on a spectrum analyzer. The analyzer shows field strength at the clock fundamental and at its harmonics. There is also signal content at frequencies between the harmonics, due, in part, to trace or cable radiation where transmission lines are involved. The transmission lines are of different lengths, where the round-trip time is not harmonically related to the clock. Note that the radiation from an infinite transmission line carrying a single step function does not display any harmonic quality. Thus, spectrum smearing should be no surprise.

10.12 CHASSIS GROUNDS AND GROUND PLANES

A metal chassis is a perfectly good ground plane. It is usually impractical, however, to use it effectively. The ground planes in the circuit boards are also circuit common. Most of the signal transport energy is confined between the traces and the board ground plane. Even if the chassis is multiply connected to the circuit board ground plane, little signal current flows in the chassis or grounding conductors. The chassis should still be connected to the circuit common to avoid electrostatic coupling to the circuits. If a very large field is inside the enclosure, then every loop will couple to this field.

When the chassis is a bulkhead used to terminate a coaxial cable, the confinement of signal field needs to be considered. The coaxial path should be continuous through the bulkhead up to the cable termination. If the connector is bonded to the bulkhead, the bulkhead is connected to the circuit common through the coaxial sheath. This causes no problem because the signal field is confined inside the sheath. In nonanalog applications, isolating the connector from the bulkhead is usually unnecessary.

When a ribbon cable leaves a circuit board, it is best to route the cable on the chassis. This keeps the cable from coupling to internal fields in a common-mode sense. To be effective the circuit board ground plane should be connected to the chassis at the connector. To extend the ground plane correctly a strap as wide as the cable should be used. Ideally it should be bonded across its width at the ground plane and on the chassis. If this connection is not made at this point, then the ribbon cable can couple to common-mode signals in the loop formed by the ribbon cable and the nearest chassis connection.

10.13 SUSCEPTIBILITY

The techniques used to limit radiation also limit susceptibility. The designer must use judgment when it comes to many aspects of layout. Before the product is built, there is no simple way to decide how many precautions to take. When little cost is involved, the designs should follow the best practice. Making changes later is always expensive.

Not all issues are apparent at the time of design. The operation through cables to other equipment often adds a new set of difficulties. Users eventually put every piece of equipment to an ESD test. Line voltage surges or line disturbances eventually take place. Designs meeting high standards are probably not susceptible to these interferences. These difficulties make a designer think twice before cutting too many corners. Much of the discussion in this chapter is intended to show what can be done to reduce susceptibility. When design approaches seem to work, the tendency is strong to leave well enough alone.

Regulations in general focus not on operation quality but on radiation issues. It is up to the designer to build equipment that operates well in the presence of interference.

CHAPTER 11

SCREEN ROOMS

11.1 INTRODUCTION

Screen rooms are important because of the many principles that must be exercised in design and application. A discussion of these principles is helpful for understanding how fields are controlled. It is hoped the reader will acquire some additional insight by reading about this branch of engineering.

The use of the word *screen* dates back to when wire mesh was used to build an electrically quiet region. A quiet room might be used to make radiation or noise measurements without interference from other local radiators. To meet the demands for more and more field attenuation, the rooms that are available commercially are built of solid conducting walls. Screened rooms can be used to keep radiation from escaping or to keep interference from entering.

11.2 SCREEN ROOM DESIGN

A screened room must have no apertures. It makes little sense to screen most of a room yet leave an opening. Every window, door, or ventilation path must be specially treated. Every conductor that enters the room must be filtered. The power line filter must be bonded to a wall of the room. The equipment ground conductor must bond to the outer wall and be picked up again on the inside surface. A filter in this lead is not permitted by NEC Code.

Door apertures can be protected by finger stock that limits the aperture dimension to the spacing between fingers. If the finger stock and its connections are hidden inside a U channel, the depth of the channel forms a waveguide. Waveguide attenuation is necessary, or each aperture will allow field energy to enter. The U channel keeps clothes from snagging on the finger stock. If the finger stock gets torn off, the room loses its shielding effectivity.

Ventilation apertures are often shielded by honeycombs that attenuate field energy by using the principle of the waveguide below cutoff. The honeycomb must be bonded around its perimeter, or the resulting apertures will violate the integrity of the screen room. If the honeycomb is removed for cleaning, any gaskets must be handled per manufacturer's instructions.

The shielding effectivity of a screen room varies with frequency and with the nature of the field. Very thick steel is necessary to provide attenuation for power-related magnetic fields. These near-induction fields are very hard to shield. It is good practice to locate the screen room away from these fields if possible. It is unwise, for example, to run power-carrying conduit along or parallel to the outside wall of a screen room. The power entrance should be at a corner of the screen room so that the conduit can enter at right angles to the wall.

An often-asked question is, should the screen room be grounded to earth at a nearby point. Such grounding is unnecessary and probably counterproductive. Multiple grounds connected to the screen room invite building current to flow in the walls of the room. Some of this current (albeit a small amount) finds its way to the inside surface. Current on the inside surface means there is field inside the room. This current is probably power related where skin effect is not too effective.

Safety is provided by the equipment grounding conductor. No other earth connection is necessary. The question of lightning protection is then raised. If a screen room received a direct lightning strike, aside from the acoustic noise, someone inside the room would not know there had been a strike. Someone outside the room is in no more danger than he or she would be inside a steel-reinforced building. If separate grounds are not added to steel boilers, motor frames, or building steel, it makes no sense to add a ground to a screen room. If one must be added, it should connect to the equipment ground at the power entrance. This extra ground is perfectly legal. It is the neutral that cannot be regrounded.

The screen room is grounded at high frequencies by the capacitance the conducting floor has to earth or building steel. The floor can be made of two conductive layers to keep stray power currents from flowing in the inner conductive layer. Doubling the layers does not reduce the capacitance to earth.

Fields that reflect from the walls cause current to flow on the wall surfaces. This wall current must make an abrupt change in direction at the wall boundaries, essentially a quarter-turn, which change is in effect an inductance. It is easier for some of the current to enter the wall than to follow a sharp bend. If current flow gets to the inside surface, the shielding effectivity is reduced.

To overcome this difficulty the corners of the screen room are filleted on the inside. The fillet might provide a radius of 1 in. and must be bonded to the wall along the entire wall edge to be effective.

Power, signal, and telephone connections should be made in one area. This limits the surface currents that might otherwise flow. All leads need to be filtered at the bulkhead, or they will contaminate the room. The filters must bond to the outside wall of the screen room. Coax bonded to the wall provides a tunnel where signal can be transferred in and out of the room. It is up to the user to maintain the coaxial integrity of the signal conductor so that it is not allowed to radiate into the room. If a signal conductor crosses the screen boundary, the conductor is a two-way street. The user has the responsibility to make sure that the signal line does not couple to external field energy and again contaminate the room.

Ducts carrying air to the room are usually insulated from the room to limit ground current flow. The insulating section should be at least 6 ft long. This precaution is taken to avoid any side flashes from lightning.

Holes drilled in the screen room wall are apertures. If they are extended in length by a bonded conducting tube, the entrance becomes a waveguide and poses no threat. If a single conductor threads the tube, this entrance is now coax and good to dc, which is much worse than the hole by itself. Threading a conductor through a honeycomb structure violates the entire structure.

Data can be sent over fiber optics. The fiber is nonconductive and does not couple to field energy. If this fiber is brought in through a conducting tube (waveguide), the screen room is not contaminated. If a steel support wire is left on the fiber-optic cable, the room is violated. A nylon support may have to be used to enter the screen room. The steel support wire on an optics cable should not be carried to within 6 ft of the screen room. This cable can carry a lightning pulse which can flash over to points in the facility. The steel should be bonded to the grounding electrode system where it enters the facility, to avoid this difficulty.

11.3 SCREEN ROOM EQUIPMENT

When radiation tests are performed, the only equipment in the room should be the device under test and any receiving antenna. If other equipment is required, the radiation from this equipment should be checked before it is used.

Lighting brought into the screen room must be of an approved type. Incandescent lighting is usually supplied for room light because fluorescent lighting does radiate. Telephones brought into the room must be filtered at the bulkhead to avoid bringing in fields.

11.4 SCREEN ROOMS FOR SECURE OPERATIONS

The tables are often turned on a screen room. The data being processed are secret, and radiation from the room is the issue. The power line filter is usually able to function in this direction, although it should be electrically located inside the room.

Experiments and operating equipment should be kept away from the room walls. Near-induction fields are difficult to attenuate, and distance to the screen wall is in the right direction. Any cables brought into the room must have their shields bonded at the bulkhead. If the termination occurs inside the room, the fields in the room will find a path to the outside world.

In a few instances entire buildings have been made secure by adding a metal skin to the structure. It is very easy for these facilities to be violated by workers not familiar with the nature of radiation. A hole for telephone access is an aperture. The telephone cables themselves will couple fields from the inside to the outside world unless the cables are in conduit that is properly bonded at the interface. The lines themselves must be filtered, or field energy will be transported out of the protected area on these lines.

A screen room is a six-sided affair. Ignoring the floor as one side of the enclosure makes the room ineffective. The earth itself is not a good enough conductor to be used for shielding.

Grounding conductors that exit a secure facility carry field energy out of the facility. This conductor radiates field since it is a horizontal antenna. If the grounding conductor terminates on the screen room enclosure, such radiation does not occur.

CHAPTER 12

ELECTROSTATIC DISCHARGE

12.1 INTRODUCTION

Electrostatic discharge (ESD) is a common occurrence. It occurs in a dryer tumbling clothes; it occurs in dry weather when hair is combed or when shoes rub against a rug. In a more dangerous environment it occurs in silos when grain is loaded. Here a discharge can result in an explosion. Friction removes charge from one body and delivers it to another. This accumulation of charge can cause potential differences that can arc over considerable distances. It occurs as lightning between clouds and from clouds to earth.

A very serious threat of damage occurs in the handling of semiconductors. If a discharge takes place so that the current path uses the silicon structure, then hidden damage can occur. This is why a great deal of care is taken in manufacturing environments to avoid any electrostatic buildup. A few microjoules of energy is all it takes to do damage. Sensitive components are shipped in conductive bags, for example, to keep motion during shipping from developing charge.

Grounding in itself is not sufficient to control damage from ESD. When an insulator builds up a charge, contact or proximity to a grounded conductor invites a discharge. If the conductor were floating, the discharge might not occur. Grounding simply keeps all of the interconnected conductors at the same potential. It is hard to ground everything. Items being unpacked from plastic wrapping can build up a charge and immediately discharge to a nearby grounded conductor.

The best protection against ESD is humidity control. When the humidity is above 40%, the probability of charge buildup is limited. Adding humidifiers to a facility must not leave pockets of dry air. All of the air must circulate through the humidifiers, or the threat of ESD still exists. Another protection is slightly ionized air. Any flame, such as a candle or match, will limit charge buildup in the local area. Ion-generating devices are commercially available.

12.2 ELECTRICAL NATURE OF ESD

When a person picks up a charge on a rug and reaches for a doorknob, the body charge moves to the fingertip. The electric field concentrates between the fingertip and the knob. The field energy per unit volume is proportional to E^2. This means that 90% of the energy is stored in the field in a radius of 1 ft. When the discharge takes place, it takes about 1 ns for field energy to travel 1 ft. For a typical pulse the current might rise to 5 A in 1 ns. By the ideas of Section 5.5, an ESD pulse is characterized by a 5-A signal at 300 MHz.

The energy released during discharge heats the air and radiates into space. The amount of energy that radiates is a function of the voltage just before breakdown. Maximum radiation occurs at about 6 kV. Above this voltage the energy that goes into heat and sound increases.

12.3 ESD TESTING

A variety of ESD testing devices (guns or zappers) is available. These devices generate an ESD pulse at a set voltage and repetition rate and are used to stress equipment in a controlled manner. Such testing is advisable for all electronic equipment sold commercially, particularly digital hardware. Malfunctions and/or destruction of equipment during dry weather can be very costly. In medical electronics such testing is critical.

ESD testing should start at a low voltage and involve points likely to be touched by an operator. A matrix of test points and voltages should be planned. The voltage is raised in steps. At the first sign of trouble the test is stopped, and improvements in design should be made. The test should progress through all voltages and should not jump from 1 to 12 kV, for example. If the device passes this test, it probably will not radiate since susceptibility and radiation are closely linked phenomena.

ESD testers can be used in two modes. In one mode the ESD arc is at the test probe. In the second mode the probe is in conductive contact with the DUT and the arc takes place in an internal gap. This second mode generates a current pulse which does not duplicate an actual ESD pulse. In this second mode, field energy is not present to enter nearby apertures. This type of test can help in separating variables when the entry mechanism is not understood.

12.4 HANDLING CHARGE ACCUMULATIONS

In most situations the accumulation of charge develops a potential difference to earth. The earth is the zero-reference conductor for all electronics on earth. If the device is battery operated or powered through a transformer, there may be no simple path for discharging a charge accumulation.

Consider a hand-held telephone powered through a transformer where a floating conductive shield has been placed around the circuitry. If a person with a built-up charge touches the handset, some of this charge moves to the handset. During ESD testing there may also be a charge buildup. An accumulation of charge means there is a voltage on the handset between the shield, the circuit, and the primary power conductors which are grounded. If the voltage is high enough, arcing occurs across the transformer.

The arcing path uses the shield to reach the transformer. In some cases the shortest path is through the shield to the circuit, which is then connected to the transformer. If this path is through a semiconductor, the circuit will probably be destroyed.

The proper way to treat this problem is to bond the shield to the secondary circuitry so that the lowest-impedance discharge path avoids the circuitry proper. A bond to the circuit common at the wrong point invites a discharge path which can also destroy components.

A conductive path from the circuitry and shield to the primary need only be 100 megohms to limit charge buildup. If the shield has 100 pF of capacitance to earth, the resulting time constant is 0.01 s. If a conductive path is not provided around the transformer, repeated arcing through the transformer may do some damage.

Paint-spray equipment poses a special problem. If the equipment ground is lifted, there may be charge buildup as the paint moves through the nozzle. Paint vapor is highly volatile, and an arc can set off an explosion. To guarantee a discharge path to avoid charge buildup, a 100-megohm resistor can connect circuit common to the primary neutral. The neutral is earthed and is a discharge path for charge buildup. This solution does not address other safety issues which require that the equipment ground be connected to all exposed metal parts.

APPENDIX

INTERFERENCE CASES AND THEIR SOLUTIONS

THE TITANIUM MELT

Titanium is melted in a crucible inside a vacuum chamber. The heating involves a controlled electron beam of 40 A at 6000 V. The beam is controlled by a computer. If the beam malfunctions, it is turned off. Impurities in the melt can boil out and intercept the beam. Often, the result is that the beam diverts to the chamber wall. If it is not turned off immediately, the beam burns a hole in the chamber. It might take a month to get back in business.

The problem is simple. When the beam diverts, the computers crash and the beam is no longer controlled. The designers blamed the problem on grounding and asked users to ground the computer and the crucible to separate "clean" grounds. All attempts along these lines were unsuccessful, and the product was in trouble.

The fix was to make sure that the return path for the current beam was kept coaxial with the beam during normal and diverted modes. The beam current was being returned through an undefined grounding path. The field created by the large loop area was large. During transitions the radiated energy was significant. Controlling the return path stopped the radiation, and the computers no longer malfunctioned. Separate (illegal) isolated grounding was removed. A pseudo ground plane was built by using the racks of hardware and the crucible bonded together.

THE THIN-FILM FACILITY

A thin plastic film was passed over rollers. If a tear or too many holes appeared, the process was stopped. The old system used fingers that contacted the roller when a hole passed by. The current flow in the rollers was sensed and used to ring an alarm.

In a facility upgrade a computer was installed. Management would not allow the computer to take over unless it could be demonstrated to work. Every time the fingers detected a fault, the computer malfunctioned. The fingers could not be disconnected until the computer could take over.

The circuit to the fingers used the facility ground for a return conductor. When the voltage to the fingers was supplied coaxially, even though the roller was grounded, the problem went away. The undefined return path radiated energy that upset the computer.

BURNING UP INTERFACE CARDS

Two computers were installed side by side in a facility. The computers were connected together over a data bus. Approximately once a month one of the interface cards would have to be replaced.

The facility was built with isolated grounds. This raised the impedance in the equipment ground path, limiting the action of line filters in each computer. The solution to this problem was to eliminate the isolated grounds by connecting equipment grounds, conduit, and receptacles together for both computers.

THE TWO SPOTS

An ion beam was used to deposit conductive material. The beam was controlled by a computer in a servo loop. In some instances the beam seemed to split and focus at two points.

A separate equipment grounding path had been provided for the product being plated. This path was inductive and added a degree of freedom to the servo system. In effect, the beam positioning servo was oscillating between two limits. To fix the problem the equipment ground was returned coaxially with the beam. The inductance was reduced, and the system was then stable.

THE WALL-MOUNTED SOLENOID

A sensor was detecting mechanical position and sending a signal to a logic circuit 10 ft away. The sensor and circuit were connected by a braided cable. Whenever the solenoid operated, the logic circuitry reacted and caused a malfunction. The cable contained power supply lines and signal lines.

The engineer had added a conductor in the cable that connected the two circuit boxes together. This conductor paralleled the braided shield that also connected the two boxes together. This added conductor was, in effect, an extra drain wire. The drain wire carried conducted noise current next to the logic signal lines and the signal and noise conductors were coupled together. Stated another way the signal and noise fields shared the same space.

The solution was simple. Cut this added conductor and allow the braided shield to be the only connection.

ERROR COUNTERS

A special interface box was added to a digital tape recorder. The equipment had to pass a chattering relay test. This is a buzzer-type circuit wrapped around the interconnecting cable. An error counter in the recorder indicated when the message being returned did not agree with that being sent. At the beginning of the test, the error buffer overflowed in just a few seconds. The interconnecting cable was braid-shielded, terminating into typical wall-mounted connectors.

The shield was carried through the connector and bonded to the inside box wall. When the termination was made outside the box, the error counter overflowed in 5 s instead of 3 s. When the pigtail loop was removed, the time went to 7 s. When the braid itself was connected to the four connector mounting screws, the time increased to 20 s. Only after back-shell connectors were installed were no significant errors encountered. The braid was not an optimum cable, but in this case the field energy was entering the system at the cable terminations.

THE NOISY COPIER

A medical product consisted of a sensor and an amplifier. The sensor was connected to the amplifier over a short shielded cable. When the ozalid printer was turned on, the device failed. The engineer had tried an isolation transformer, a ground plane, and various earth grounds. The number of clip leads on the bench indicated the level of his frustration.

The fix was simple. The shielded cable was simple coax. When a second braid was added over the first shield, the noise problem vanished. This shield was bonded at the amplifier common. Noise currents no longer flowed in one of the signal conductors.

LIGHTNING AND THE BIG FACILITY

In the Southwest desert, summer thunderstorms are common. A large facility had been built with a single-point facility ground located several hundred yards from each building. This was the ground point for all signals processed in the facility. The problem was that during lightning activity equipment was being blown up. A large newly installed isolation transformer had been destroyed by lightning.

Lightning hitting the earth anywhere in the vicinity can easily produce a ground potential difference of 15,000 V between different facility grounds. If the primary coil of every transformer is grounded by the neutral at the service entrance, this 15,000 V appears across every primary and secondary coil. The rated stress level is usually 5000 V. This method of grounding is not valid for a large complex.

The preferred solution is to eliminate the central grounding philosophy. Signal lines between buildings must be isolated and protected individually. When lightning hits, any excess current is diverted to earth via surge suppressors and gaps. It is now possible to handle data transmission over optical links and avoid this problem completely.

20 AMPERES IN BUILDING STEEL

In a high-security facility all of the grounds were carefully routed to a central point. It was felt that this would keep anyone from sensing this current and limit a possible security leak. This approach has many deficiencies, but, aside from this fact, the engineers were very upset when it was discovered there were 20 A of power current flowing in one steel beam in the facility. This meant to them that the current flow paths provided were not fulfilling their role.

An examination of the facility showed that there were two large distribution transformers mounted on a steel structure on the third floor. The steel structure near the transformer formed loops and the transformer's leakage field was causing current flow in these loops. This current found its way through the entire facility.

The solution was to insulate the transformer mounting to eliminate the nearest loop. This reduced the current by a factor of 10. Any further reduction would have required moving the transformers. At a lower level other mechanisms would then have to be considered.

THE LIVE MISSILE

A facility had been shut down because a missile had somehow become armed. During shutdown a facility inspection was made. The inspection brought to light improvements needed in the lightning protection system. The sides of the building involved corrugated steel panels. These panels would function as down conductors providing the lowest-impedance path for lightning to follow. Unfortunately the panels were not terminated at ground level. Lightning using this path would arc at ground level generating a large electromagnetic field. This arcing point would be a few feet away from active but unarmed missiles.

The fix for this problem was simple: provide a series of grounding rods that would earth the panels properly. Bond these rods to the ground ring that had been provided.

WAVY LINES

A graphics facility uses large high-resolution display terminals to demonstrate its work. Any anomaly in the display is unacceptable. The disturbing aspect of the display was a raster that was not square. The right and left edges were noisy, and this noise seemed to be power related.

An examination of the power distribution system indicated that isolated grounds were used, which added impedance to the equipment grounding path. Filters associated with the display terminals were limited in their effectivity because of this impedance. When the video cables were connected to a computer, some of the filtered current would use the video link to return current via the computer's equipment grounding path. This path coupled noise to the video signal.

The fix was simple. Eliminate the isolated grounds in that facility. The equipment grounds were bonded to the receptacle and conduit in every outlet. With this fix the noise was not discernible. Plans to install a computer floor were shelved.

OSCILLATION

An order for 20 instrument amplifiers was returned to the manufacturer to limit the voltage output to 10 V. Zener diodes were installed on an internal output circuit that had its own current protection. Upon retesting it was determined that some of the gains had shifted by more than 0.1%.

The zeners were not loading the circuit because they were not conducting. The trouble was traced to the capacitance of the zener diode. This small reactive load was sufficient to cause the internal stage to oscillate at a frequency well above the band of the instrument and of the instrument's output stage. For this reason the instability was not immediately noticed.

The fix was as follows. A resistance of 20 ohms was placed in series with the zener diode. The voltage was still limited to 10 V and the circuit was still stable. It would have been incorrect to change all the gains to accommodate the instability.

PLASTIC CIRCUIT BOX

A piece of telephone-related equipment was undergoing ESD testing. This device was power-operated with a small power transformer. The inside of the case was coated with a conductive paint. After the equipment was pulsed six times, the seventh pulse would destroy the device.

The problem related to the fact that charge was being built up on the entire device and there was no drain for it. Each ESD pulse added to the charge accumulation. The capacitances between the primary and secondary coils of the transformer and between the painted shield and the circuit were involved in a voltage divider. When the charge accumulation was sufficient, the resulting voltage would break down transformer insulation. The resulting arc would flow from the shield to the circuit and across the transformer coils and destroy the circuit.

It is not a good idea to accept arcing in the transformer as a solution because pinholes can eventually cause breakdown. It is preferred to provide a drain path around the transformer so that there is no charge buildup. A 100-megohm resistor is acceptable. Another solution involves an MOV between the primary and secondary coils that conducts above 500 V.

It is desirable to connect the painted shield to the circuit. If no connection is made, a breakdown must occur to the circuit, and the conductive path provided around the transformer will offer little help. This shield connection should be made at the transformer so that any pulse of current that flows does not flow in the circuit proper. Again a 100-megohm resistor can be used to drain off any charge accumulation on an ungrounded shield. An MOV can be used to limit voltage if the resistance is not permitted.

COAL-BURNING TEST

An experiment in obtaining energy from powdered coal was in trouble. The control system was receiving false signals, and as a result the system could not be made operational. The engineer added filters to the power supply and to the control leads and the problem was not fixed. When filters were added to the output sensing leads, the problem was solved. Every lead entering a test box can carry interference. Filtering some of the leads is not sufficient.

A PRODUCT DISTRIBUTION CENTER

In a large city a building covering an entire city block handles orders from many supermarkets. Goods are brought in by truck and placed on palettes. Interconnecting the palette stacks are miles of conveyer belts that can transport boxes of product. A central computer produces bar codes that are affixed to the boxes before they are placed on the conveyer belts. Through a system of conveyer shunts an order for many different items from a particular market is stacked at a pickup point for truck loading.

The system was so efficient that, even though it made errors, management refused to allow anyone to turn the system off. The trouble resulted from fields generated by the many solenoids and motor controllers that operated throughout the facility.

The computer was inside a metal box, and all of the control leads were optically isolated. The computer was located in the center of the facility right in the middle of significant interference. All control leads were brought into the box containing the computer. Optical isolation occurred on integrated circuit boards located next to the computer circuitry.

The fix involved moving the optical isolation to the bulkhead interface. This kept the fields coupled to the control leads from entering the computer.

INDEX